Heraclitus Redux

Collective Studies in Knowledge and Society

Series editor:

James H. Collier is Associate Professor of Science and Technology in Society at Virginia Tech.

This is an interdisciplinary series published in collaboration with the Social Epistemology Review and Reply Collective. It addresses questions arising from understanding knowledge as constituted by, and constitutive of, existing, dynamic, and governable social relations.

The Future of Social Epistemology: A Collective Vision
edited by James H. Collier

Social Epistemology and Technology: Toward Public Self-Awareness Regarding Technological Mediation
edited by Frank Scalambrino

Socrates Tenured: The Institutions of 21st Century Philosophy
Adam Briggle and Robert Frodeman

Social Epistemology and Epistemic Agency
edited by Patrick J. Reider

Democratic Problem-Solving: Dialogues in Social Epistemology
Justin Cruickshank and Raphael Sassower

The Kuhnian Image of Science: Time for a Decisive Transformation?
edited by Moti Mizrahi

Taking Conspiracy Theories Seriously
edited by M. R. X. Dentith

Overcoming Epistemic Injustice: Social and Psychological Perspectives
edited by Benjamin R. Sherman and Stacey Goguen

Heraclitus Redux: Technological Infrastructures and Scientific Change
Joseph C. Pitt

Minority Report: Dissent and Diversity in Science
William T. Lynch (forthcoming)

Heraclitus Redux

Technological Infrastructures and Scientific Change

Joseph C. Pitt

ROWMAN &
LITTLEFIELD
INTERNATIONAL
London • New York

Published by Rowman & Littlefield International Ltd
6 Tinworth Street, London SE11 5AL
www.rowmaninternational.com

Rowman & Littlefield International Ltd.is an affiliate of Rowman & Littlefield
4501 Forbes Boulevard, Suite 200, Lanham, Maryland 20706, USA
With additional offices in Boulder, New York, Toronto (Canada), and Plymouth (UK)
www.rowman.com

British Library Cataloguing in Publication Data
A catalogue record for this book is available from the British Library

ISBN: HB 978-1-78661-235-9

Library of Congress Control Number: 2019949869

Dedicated to
My wife, Donna,
And
Joseph Oliver Falkinham, III

Contents

Preface

This book has a long history. It started out some 10 years ago as a rather straightforward review of standard philosophical accounts of scientific change. It was to end with a novel account of scientific change wherein the role of technological innovation was introduced into the discussion, arguing that in many cases scientific change, meaning by that theory change, was a direct function of the development of a new technology. The germ of that idea was placed in my head by Joe Falkinham, a microbiologist with whom I taught a course on humanities and the biological sciences for about a decade. In casual conversation one day he remarked on the difficulty people using the electron microscope when it first came on the scene had in understanding what they were seeing. Sometime later I wrote a small book, *Thinking About Technology; Foundations of the Philosophy of Technology*, in which I introduced the idea of the technological infrastructure of science. The technological infrastructure is what makes the doing of modern science possible, all the instruments, labs, funding agencies, and so on.

The concept of a technological infrastructure began, over the years, to color how I viewed, first, the relationship between science and technology and then, second, how I came to understand how various technologies enable virtually everything we do. As has been widely noted, the pace of technological innovation is accelerating and the younger generations are totally tied in.

But it wasn't until my small rural community of Newport, Virginia, was threatened with destruction that I fully understood how useful this idea was. In September 2015 we were informed that a 48-inch diameter pipeline was going to be coming through not only our farm, but significant parts of the small town we live outside of. It was to be used to carry fracked gas from the Marcellus Shale gas fields in northwestern West Virginia through West Virginia down and then across the Appalachin Mountains into Virginia and east to connect with a north-south pipeline and eventually to four East Coast ports where it was to be processed and shipped abroad. When we started trying to figure out how to stop it was when we became aware of the massive infrastructure behind it. Funding sources, various transportation systems, mining companies, government agencies, politicians, and more combined to overwhelm the actions of 100 country folk. But seeing all of this as a technological infrastructure helped me, at least, understand how it worked and where the pressure points were.

This realization made me rethink the ultimate focus of this book. Yes, I still wanted to work on scientific change, but that was going to be used to set the stage for a broader societal critique. That is what I hope I have achieved. I hope I have developed a means by which we can understand the multifaceted roles technologies play in our lives and what we can do to have them help us in our various endeavors.

None of this would have happened without the assistance of many people. First and foremost is my constant inspiration and support, my wife Donna. She helps make things happen. She is patient and understanding and my best proofreader. In addition, there are my colleagues here at Virginia Tech who have supported me and allowed me to develop my work in the philosophy of technology. There are many members of the Society for Philosophy and Technology, Paul Durbin, Davis Baird, Carl Mitchem, Deborah Johnson, Diane Michelfelder, and many more. There are a number of people who have been helpful and supportive of me over the years. Chief among these is Nicholas Rescher. A special thanks to Ashley Shew and Ron Laymon who both read an earlier version and gave me incredibly helpful extensive comments. James Collier for his support and encouragement and Terry Zapata and Holly Belcher. Finally, my thanks to my students. Also I wish to thank Diana Hoyt for suggesting the book's title. Virtually everything I have written has come from their enthusiasm and lust for learning.

Chapter 1

Introduction

> Everything changes and nothing remains still. . . . and . . . you cannot step twice into the same stream.
>
> (Heraclitus, as quoted in Plato's *Cratylus*, 403)

This book was initially about only scientific change. While scientific change remains one of its foci, to treat that topic more fully I have had to step back and examine science also in terms of the social phenomenon it is. I begin with the assumption that what we call "science" is a set of multifaceted social processes in a constant state of change. Accepting this as our starting point has significant ramifications both for how we do the history and the philosophy of science and technology and for some previous philosophical attempts to come to grips with scientific change. Ultimately, however, the analysis produced here not only offers a different take on how to understand scientific change but also provides us with a mechanism to understand how technologies influence what we do in general.

Some preliminary considerations need to be brought to the fore. First, a commentary on the previous quote from Heraclitus. First of all, why Heraclitus? Heraclitus's comments stand in stark contrast to Plato's view of reality, a view that continues to haunt us and may be remotely responsible for certain assumptions we hold about the nature of the universe and what we can know about it. Thus, for Plato, what we experience on a daily basis is not reality. It can't be real because it suffers from growth and decay—in short, it is in a constant state of change. Reality, on the other hand, is perfect, which means, for Plato, that it is permanent and unchanging. It is the world of the forms, which exist in a different realm. Things on earth are what they are by virtue of participating in a form of the thing. For example, a chair is a chair by virtue of participating in the form of chairness. Heraclitus's view contrasts deeply with Plato's. But it is in accord with what we know about the universe, that is, that it is changing.

When Heraclitus says, "Everything changes and nothing remains still," he could mean one of two things: (1) everything we know changes or (2) everything changes: what we know, the world, the universe, everything. It is obvious that everything we know changes; we are constantly discovering new things, new arrangements of things and people, and so on. The harder position to defend is the claim that everything changes, not only what we know, but the universe itself, for this proposal challenges the cherished idea that, among other things, there are unchanging universal physical laws governing the universe. But we have some license given Heraclitus's reference to the changing river. And if we combine the two, that both what we know and that the universe change, things get messy in a hurry. There is a trivial sense that what we know changes because the universe changes. But to claim that everything changes, independently of what we know, poses some difficult problems. Nevertheless, if we stop to think about it, it is true.

The universe began, we are told, with a big bang—the original elements were helium and hydrogen and as of this time we have 118 elements and physicists think they have discovered yet another that also brings with it a fifth force. The universe itself is in a constant state of change, expanding, novas, colliding galaxies, and so on. Evolution keeps the biology of earth in constant flux. Even the geology of the earth is in constant motion. And, of course, as we find these things out our knowledge changes.

Second, science is not some monolithic phenomenon. It is not a thing; rather, it is a collection of many different activities in which people delve into the multiple dimensions of physical reality. The assumption that science is somehow one thing has been fed by claims of, among other things, *a* scientific method. It is a standard claim: we know biology is a science because it uses the scientific method. But with just a little reflection we can see that there are many methods employed by the various sciences. Further, each of general headings we use to identify a science is just that, a general heading. Just as there is no scientific method, there is nothing called Biology. There are many biological sciences from microbiology to ecology, and they employ different methods and technologies. One of the big problems within any of the scientific categories is to see how developments in one area bear on the other areas or not. This is true for all the so-called sciences. What makes something a science is hard to identify once we get rid of the scientific method. One stab at identifying *the sciences* is to see them as a collection of activities in which the practitioners are attempting to understand how the world works and how it came to be the way it is and, possibly, what its future may look like. Each science looks at a different part of the world; cosmology examines the heavens, nanoscience has as its focus the very small. There is no one science that puts it all together—that is the job of the philosopher, as Wilfrid Sellars noted so well.

> The aim of philosophy, abstractly formulated, is to understand how things in the broadest possible sense of the term hang together in the broadest possible sense of the term. Under "things in the broadest possible sense" I include such radically different items as not only "cabbages and kings," but numbers and duties, possibilities and finger snaps, aesthetic experience and death. To achieve success in philosophy would be, to use a contemporary turn of phrase, to "know one's way around" with respect to all these things, not in that unreflective way in which the centipede of the story knew its way around before it faced the question, "how do I walk?," but in that reflective way which means that no intellectual holds are barred. (Sellars 1963, p. 1)

Third, our initial assumption that science is constantly changing forces us to consider science first and foremost as a historical entity since it is to be understood as a set of processes, and since processes occur over time, the sciences, therefore, should be understood as historical processes. Taking this approach makes it easier to see the sciences as human activities embedded in a world of many human activities.

Fourth, in treating the sciences as historical processes we must abandon our philosophical hopes of universal answers to alleged universal questions. More specifically, this chapter focuses on some of the historical sources of change and the implications for traditional philosophy of science of determining how things actually proceed in the sciences.[1] An account of scientific change should not be about abstract criteria for choosing among competing theories in the event of an entrenched theory's failure, primarily because that rarely, if ever, happens. It is not about being rational or the alleged rationality of science. To pursue those approaches is to be fundamentally ahistorical. In what follows I especially eschew more traditional philosophical approaches to understanding the problem of scientific change as a problem of rational choice, ala Popper, Kuhn, Lakatos, and Laudan.

The basic claim of the position argued for here is that often scientists seek new theoretical frameworks when new technologies[2] allow them to "see" what could not be seen previously. I am not talking about a Kuhnian paradigm shift. Rather, I am concerned with the prior question of why paradigms or theories are challenged to begin with. Does something happen that "forces" the change? As with everything else, there is no universal answer to that question. Some theories are falsified.[3] Some theories are abandoned because they no longer do the job they were supposed to or other theories that do the job better come into play. Some theories are abandoned because they no longer can be used to explain the phenomena under investigation. The cases I am interested in fall mainly within this third rubric. In these situations, it is often the case that the explanatory failure is due to the fact that a new technology, such as Galileo's telescope, allows us to see what could not be seen before and for which new phenomena we have no current explanation, for example, Galileo's discovery of the moons of Jupiter. Explanatory failure is not the only source of interest here. Sometimes new technologies make possible what would not have been possible without them, such as a new social technology like the National Science Foundation that encourages new research that could not have been undertaken without its funding resources or the electron microscope that opened the world of the very small.

The one major philosophical consequence of this enquiry, that some significant change in science is a function of new and innovative technologies, is that some basic notions we associate with science like "evidence," "observation," and "explanation" also change over time and are science specific. This realization should force us to reconsider attempts to answer such questions as "What is scientific evidence?" while ignoring the historical and social situations in which what was claimed as evidence was used. For example, Galileo's claim, that with his telescope he had evidence that there was more than one center in the universe about which celestial objects rotated, was challenged by denying that such "observations" could be used as evidence. One claim was that it was impossible to see something accurately in the perfect heavens using a device made on earth, which, by definition, means it is imperfect (echoes of Platonism?).

In short, what I hope to do here is to force a reconsideration of philosophy's doomed search for eternal answers to what have been considered eternal questions, or what Wilfrid Sellars and, before him, Leibniz[4] called "the Perennial Philosophy" (Sellars 1963, Chapter 1).[5] Genuine philosophical questions are not eternal, they are of the moment, and so must be their answers.

To take this approach is not to say that concerns over what constitutes evidence are not legitimate; it is rather that "evidence" is, at best, the name of the category, the criteria for inclusion in which changes over time, likewise "observation," "explanation," possibly "theory." To ask what counts as scientific evidence is necessarily to ask what counts at-this-time-and-place-for-this-science. Thus, to take on the topic of scientific change requires that we approach the sciences as the historical processes they are, not as some abstract metaphysical thing.

The view I am championing has implications beyond the philosophy of science. It extends to all areas of philosophy. One way to see the import of what I will be arguing over the course of this chapter is to consider what we expect of a satisfactory answer to a philosophical question. There is a view, sometimes referred to as the Perennial Philosophy, that claims that a philosophical question is a universal question for all time, hence Perennial. In the

Perenial Philosophy it is also assumed that there is one response that answers the question for all time and the philosopher's task is to find that answer. But that assumes that, because the words of the questions remain stable over time (translation from non-English languages notwithstanding), the content of the question remains the same. But the content, that is, the meaning, of the questions does not remain the same.[6] To see that simply consider what would be accepted as an adequate answer at one time as opposed to a later time. An acceptable answer to "What is the Good Life?" for the Athenians of the fifth century BCE differs from the answer we expect today. In a society where women did not have the vote and slavery was tolerated, the conception of the Good Life must have been different from what we expect today when slavery is viewed as an abomination and we are still exploring what it means for women and men to be equal. That we continue to assume the ancient Greeks had some fundamental insight into what the questions of philosophy are needs to be challenged, and their understanding of how to answer those questions needs to be questioned and ultimately rejected.

To emphasize the point that philosophical questions do not have perennial answers let us consider the nature of knowledge. Beginning with the ancient Greeks, knowledge was assumed to be developed and held by *individuals*. The philosophical emphasis was on what was required for *me* to know something. How do *I* acquire knowledge? The result was a succession of accounts in which the hopeful knower was stuck inside her head. The problem was to get from the sensory input you acquire when working in the world to some way to prove something about the world. But how to get from what you formulate inside your head out to the world has evaded satisfactory solution.

It wasn't until the 19th century that there was a shift away from the failed project to find out how I get to know something to the promising investigation of how a *community of inquirers* produces knowledge. The move was initiated by the American founder of pragmatism, C. S. Peirce. Not only does the nature of the question change, but necessarily so does the character of the answer. If the question changes to "How is knowledge produced?" the door is opened even further to empirical factors such as groups of investigators chasing a problem or a set of related problems, labs and how they work, who else is involved in the enterprise. The nature of questions also changes to deal with the question of who is in the group, what constitutes class membership, and the role of technologies. I am defending the view that contemporary science wouldn't be possible under the old epistemological concerns. Contemporary science is what the historian of science Derek Price called Big Science. Big Science is an elaborate and complicated enterprise consisting of scientists, graduate students, labs, funding agencies, institutions to house the enterprise, venues for publication, lots of publications, editors, reviewers, and lots and lots of technology. Scientific knowledge is the product of this enterprise and is a function of the group of enquirers establishing criteria for accepting proposed knowledge into the house of knowledge. These criteria change over time, especially if they are technology dependent, since the technologies change over time. The occupants of the house of knowledge also change as new discoveries cast doubt on old claims. This is not to say that there still aren't questions about how I, the individual, get the knowledge I need to get around in the world successfully. Here again, pragmatism comes to the rescue. For the pragmatist, the mark of knowledge is successful action. Successful action in the world by individuals requires what I call common sense pragmatism. This is the process of learning from experience and revising false assumptions (more on this later).

In part, some of the challenges facing philosophy today, that is, charges of irrelevance and navel gazing, are due to a continuing dedication by many philosophers to an ancient and revered conception of philosophy as a particular type of dialogue. The participants in this envisioned dialogue are other philosophers, past and present. If this is so, then Plato did indeed set the agenda, for as Whitehead noted,

> The safest general characterization of the European philosophical tradition is that it consists of a series of footnotes to Plato. I do not mean the systematic scheme of thought which scholars have doubtfully extracted from his writings. I allude to the wealth of general ideas scattered through them. (Whitehead 1929, 1979, p. 39)[7]

One way of telling this revered story is that the so-called dialogue began when Aristotle attacked his former teacher's views and the battle was on. An interpretation of what followed is that the history of Western philosophy is the history of the dialogue between Platonists and Aristotelians up to Kant in the 19th century who tried to integrate the two schools of thought. Whether that is a correct view is not the issue—it is common enough and it is also a handy way of explaining to undergraduates why they are studying this stuff.

But to see philosophy as a dialogue in some version of the view sketched earlier is to see it as a strange sort of dialogue, for, if that view is correct, that dialogue must in some sense be frozen in time. If the questions of philosophy are those introduced by Plato and if the conversation is between his followers and Aristotle, what about the rest of us? More to the point, what about the influence of the changing world in which we live on these issues as we understand them today? The frozen-dialogue view also has the disadvantage of not easily fitting new areas of philosophical inquiry into the dialogue schema. Thus, for example, where does the philosophy of technology fit? While there are social and ethical issues associated with new technological developments, it is not merely a subfield of ethics. And while technological innovation creates new epistemological issues, it does not fit easily into a standard epistemological framework, and so on. So, what to do?

Good philosophy is not frozen in time. We wrestle with philosophical problems not primarily because Plato worried about this or that, but because they arise out of contemporary concerns. When we ask "What is knowledge?" today, we do so while being aware of—but more often than not not overtly willing to address—the enormous technological infrastructure in which scientific activity, which provides us with so much new knowledge (but not all of it), is embedded (see chapter 4). Thus, it is not just what does it mean to see something with the naked eye that interests us, but it is also the question of what is the meaning of "see" when we "see" a distant galaxy by way of a "picture" taken by instruments aboard the Hubble Telescope then transmitted to earth by a system controlled by computers and computing programs. There are clearly two meanings of "see" at work here, the common sense one and a contemporary scientific one—how do we reconcile them, or do we?

The job of reconciliation was a task Sellars set for himself.[8] I am not so sure it is a job I want to pursue, primarily because if we take the problem of reconciliation as one of our primary jobs, it will never be completed, since it is permanently ongoing as scientific practice and discovery continually change and what we think we know changes. The need to reconcile new and emerging scientific knowledge with everyday common sense never ends. Even more troublesome is the fact that we cannot anticipate what our technologically enhanced sciences will turn up next. If the future of our understanding of the universe is unpredictable, it is, therefore, not clear that there is one *a priori* schema that will solve the problem of reconciling scientific understanding and commonsense for all time. To seek that one schema is to continue to buy into the initial premise of the perennial philosophy, which I reject.

One reason to reject that premise is because there is no reason to accept Plato's (and before him, Parmenidies's) own assumption about the absolute permanence of reality, a reality that is not part of this world, but transcends it, in some obscure metaphysical sense. It is Plato's commitment to the static eternality of reality that, I suspect, lies behind the continuing efforts of the practitioners of the perennial philosophy to seek universal solutions to philosophical problems.[9] But there is absolutely no reason to endorse that view today. If, rather, we endorse the idea that "science is the measure of all things, of what is that it is, and of what is not that it is not" (Sellars 1963, p. 173) and we acknowledge that the sciences are constantly revealing a universe that seems to change as our ability to see it in all of its intricacies expands, then some archaic commitment to an unchanging reality seems silly and unproductive. Here we have a clash between some unfounded metaphysical assumptions and the development of increasingly sophisticated epistemological methods. So much the worse for metaphysics, maybe we should just drop it.

The fourth and final component of the Perennial Philosophy is logic. The syllogism was formalized by Aristotle and remained the paradigmatic example of good reasoning for at least 2,300 years. The syllogism remains in use today, usually in introductory logic courses. But today when we speak of logic proofs we refer to something completely different. Logic today is understood as a system composed of a vocabulary and a grammar. The grammar specifies what constitutes well-formed formulas, rules of inference, forms of proof. The vocabulary specifies the basic elements. The point of doing a proof of an argument using this system (there are many such systems) is to show that an argument is valid. The system is said to be truth preserving if from true premises one cannot derive a false conclusion. But how do you know the system is truth-preserving? Well, you construct another system that is about the first system, a metalogic, whose purpose is to show just that. This is a far cry from listing the forms of the syllogism. Furthermore, structural features of a complete system of logic can be shown to ground computer theory.

So, there you have it. Ethics and epistemology may seem to ask the same old question, but the answers they provide prove otherwise. Metaphysics is replaced by physics, and logic is a world of its own. Long live the non-perennial philosophy!

On the other hand, in rejecting the program of the perennial philosophy do we not risk slipping into some form of relativism? That would be a serious charge and fairly devastating to both philosophy and science[10] that, all things being equal, remain committed to the search for truth, whatever that may turn out to be. Whatever truth may be, it cannot degenerate into whatever a single individual believes, and that is where extreme forms of relativism seem to end up; "what may be true for me may not be true for you" is, for non-relativists, incoherent.

The solution I am proposing looks to history to block the slide to relativism. But my view of how to use history in a philosophically enhancing manner differs from what has come to be the standard view. Philosophy of science's turn to history was made popular in the 1950s by Norwood Russell Hanson (1958)[11] and in the 1960s by Thomas Kuhn (1962). Kuhn's historical approach to the development of science led many philosophers of science to adopt the case study method, thereby giving rise to the history and philosophy of science as a general field of inquiry. Elsewhere I have argued that case studies do no philosophical work (Pitt 2001). Thus, either you select a case study to illustrate a point you have already committed to or, given the case study, nothing much follows that is philosophically interesting since the problem of induction stops generalization. Thus, case studies do not seem to be philosophically useful. But this is not to say that history isn't useful.[12]

But it may be too quick to simply dismiss case studies. This is because we don't really know what constitutes a case study. "Case studies" are used in the teaching of medicine, law, Science and Technology Stdies (STS), and other disciplines. Often offering up a case study is just that: offering up a case study, the study of a particular set of events. Sometimes a case study is used to illustrate a point. But why not call that an illustration? Sometimes it is pointed to as an example, so why not call it an example? Sometimes it is simply a point of history, so why not call it history? And the beat goes on. Rather than talk of case studies I propose we explain what appealing to this particular historical episode is supposed to do.

My appeal to history takes the middle ground between what is essentially an ahistorical perennial philosophy and the world of case studies. I propose that to understand the growth and development of the sciences we look at problematics. A problematic is an issue or a set of related issues that have attracted the attention of a number of thinkers over a period of time. Given the element of time, what the issues are may change and the solutions to the problems may turn out to be unanticipated. To take this approach is to recognize that scientific problems are not addressed in splendid isolation one at a time by one individual at a time hiding in his or her basement.[13] One only need reflect on the role Mersenne played as Descartes's *de facto* corresponding secretary in fostering scientific and philosophical conversation across Europe. Likewise, Galileo was in correspondence with Kepler and Clavius (chief mathematician at the Jesuit Collegio Romano), among others. The scientists at the beginning of what we tend to call the modern period, that is, the 17th century, were in constant communication with each other, sharing problems, solutions, concerns, findings, and so on. The picture of the genius toiling in isolation is a myth. Leibniz was fully aware of what Newton was doing and vice versa.

If we turn to this rich historical community, we will get a better sense of the development of scientific ideas than if we concentrate on one thinker and what we mistakenly assume is his (and, alas, they were mainly men) problem. These problems rarely come out of nowhere occurring to just one individual; rather, they are "in the air," so to speak. The great minds of the past were part of a more general community that had a history, working on problems their predecessors handed down, changing those problems, inventing new expressions of the problems, and so on. This remains the situation today.

Now it might be objected that this is merely doing philosophy of science in context. Doing things "in context" is very popular now. However, it suffers from a serious problem: how to define "context"? I will not attempt to solve that problem. Rather, I will concentrate on what the goal of contextualization is supposed to be. That simple objective, however, is almost as obscure as a clean definition of "context." So, instead of addressing the issue head-on, I will back into it.[14]

First, let us consider the evils contextualization is supposed to avoid:

1. Whig History—a term coined by Herbert Butterfield (1931), refers to the attempt to impose current categories of analysis on past historical events.
2. Universalism: a corollary to 1—the idea that certain features of science are constant over time.

3. Modernism—the insistence that the most important developments of any epoch are science (conceived in contemporary terms) related.
4. Abstraction—the reification of key features of an historical period.
5. Internalism—the process of examining the work of a person by appeal only to his or her notes and texts without consideration of any social or external factors—falls prey to 3 or 1, since really to know is already to understand the context in which an author writes.

Assuming it is possible to avoid these evils, there remain serious dangers facing the contextualist. For what the historian concentrating on context often does, having avoided these five cardinal sins, is to concentrate on individuals[15] and to consider the influences or the consequences of these influences for their work. What this means is the following. First, given 5, all the historian has left to do is to reveal the social and intellectual factors that might be said to motivate the views expressed by the particular historical figure under discussion. For, to provide a close analysis of the work of the person in question (it must be a person to avoid 3 or 1), exposing its logic or even its content, amounts to 5. To the extent that the views of some person or other are to be considered, it is only by virtue of the perceived audience or influences. But who the influences and audiences are falls to the historian to identify since historical figures cannot be trusted to know who really influenced them or to whom they are really responding. How the historian avoids 1, 3, or 5—or how he or she knows whom to identify as the relevant audience or influences—remains something of a mystery. The problem here is fundamental, for (1) it is not enough to say "x read y," since that alone does not establish influence; (2) nor is it enough that x quotes y for that admits neither to reacting to y or even to attempting to extend y's ideas, since x may not know what really motivates him. (The contextualist has opened the door to this objection by using it to reject internalism, i.e., *tu quoque*);[16] (3) it is equally inadequate to cite who read x, for it can be the case that x was read for all the wrong reasons (c.f., the misuse of Nietzsche by the Third Reich).

The contextualist historian is then left in the position of seemingly arbitrarily identifying specific people in specific places and can only hope that the preponderance of the evidence and correlations account for what x said about y. The laudable intent of the contextualist is to show that figures do not emerge from a vacuum. The problem, however, is that there is no obvious principle of selection that guides the identification of the people or events that allegedly transform the vacuum into a social context. The result is that the figures highlighted are sometimes minor or obscure, likewise for other social factors.[17] Without a well-articulated and defended principle of selection, the attempt to construct a context is at best arbitrary, at worse it is self-serving. Why certain figures are identified as important is not clear, since all the objections used earlier with respect to x apply equally to these problems. The contextualist project seen in this light is hopelessly flawed.

If we pay too close attention to the standard justifications for contextualization, the program collapses. And yet, there is something positive in favor of each of 1–5, that is, the rejection of Whig History, Universalism, Modernism, Abstractionism, and Internalism. It is just that taken together nothing much is left. Where have we gone wrong?

It might appear that we have been led to our unhappy conclusion by concentrating on only one aspect of the contextualization of history: the individual. But the problem not only occurs when individuals are the subject of discussion. For example, an anti-Whig historian will also justifiably reject talk of "science" in the 16th century. It would seem that the reification of concepts is also problematic.

So, what is the point of contextualization? *What is appeal to context supposed to accomplish?* Minimally, a context is supposed to provide an explanatory framework for specific historical developments. The crucial mistake advocates of historical contextualization make is to give the impression that there is only one appropriate context that satisfies the explanatory role.

The writing of history is necessarily selective (more later in the book). However, the shift from individuals or activities (e.g., science and art) to context is no less selective or arbitrary, for (apologies to Nelson Goodman 1955) contexts are where you find them. Consider the variety of contexts in which Galileo could be said to have operated:

(1) The Renaissance,
(2) The Scientific Revolution,
(3) The Medicean Court,
(4) The Archimedean tradition,
(5) The Euclidean tradition,
(6) The Aristotelian tradition,
(7) The Platonic tradition,
(8) The medieval tradition,
(9) The battle between the Vatican and Italian secular states for political control of the Italian Peninsula,
(10) The Age of Exploration,
(11) The Age of Elizabeth,
(12) The 17th century,
(13) The 16th century,
(14) A personal struggle to financially support his family,
(15) The politics of the struggle between theologians and natural philosophers, and
(16) The Counter-Reformation.

You pick the relevant explanatory framework.

However, picking the relevant explanatory framework is not as difficult as it may appear—the trick lies in figuring out what you want to explain. The mistake to be avoided is to assume that there is necessarily only one explanatory framework. What is more problematic than which framework(s) to pick is what constitutes an explanatory framework, that is, what constitutes an explanation in these contexts, or to cut to the chase—what constitutes an historical explanation *simpliciter*.

This is a big question. There already exists an extensive literature on the topic. I pretend no complete account—but I can attempt a different beginning. Part of the answer will come from an answer to a slightly different question—"What do we want from an historical explanation?"—that is, what is the point? Rephrasing, it reads "why do we seek historical explanations?" Clearly, one answer to this question is, "To learn from the past." But I genuinely doubt if we seek historical explanations merely to understand how we got to where we are. We seek more—not merely an answer to the question, "How did we get here?," but sometimes also, "how can we avoid ending up in this situation in the future?" There is little doubt that we cannot produce explanations that will give definitive answers to those questions for several reasons: (1) The analogies between the past and the present are just that: analogies. Learning from the past is only as successful as the strength of the analogy between past and present. (2) There is no single fact of the matter of the past—more information is constantly surfacing, depending on what we think we need to know. The plasticity of the past is also influenced by contemporary ideology, cultural fads, and so on.

And yet the situation is not hopeless. The search for answers from the past depends as much on the perceived state of the present as on our perception of the options for the future. And it is in the latter that we find our clues to the adequacy of historical explanation.

What I have in mind here centers on the notion of a coherent story. What makes for an adequate explanation is the sense that our account of why things happened in the past hangs together with what we know proceeded and followed the events in question. Appearances to the contrary notwithstanding, this is not Whiggish. For nothing here suggests that we necessarily must see what happened in the past in a direct, causal line with the future that is our present. It therefore makes no sense to talk about the global importance of current events, theories, and so on—since that verdict awaits the future (Pitt 1992a). A relevant set of contexts can be identified in terms of the explanatory value, that is, the coherence, they contribute to the story accounting for why what happened happened.

To the extent that the failure to include certain factors can be shown to be relevant to understanding what happened after the events in question, we can

argue for expanding the set of contexts. So, an historical context is a set of factors that provide an explanatory context for an event, a person's actions or work, or a social trend. The adequacy of the context is a direct function of its ability not only to account for the event and others, in question, but also for its prior and subsequent history.

Now it might be objected that since, as I suggested earlier, everything changes, will our coherent story also change and how does that affect its efficacy? Yes, of course, the story will change. That is not the issue. The issue is whether or not it continues to provide a coherent account, given all of what we know. There is no reason why that shouldn't be the case. We are constantly adjusting our view of the world as we learn more.

It has also been suggested that this appeal to a coherent story providing a historical explanation sounds very much like inference to the best explanation. Further, it has been claimed that there are no criteria for what constitutes "best."[18] Yes and no. Yes, the objective is the best explanation we can provide for the issue in question. But no, because I have provided a criterion: no loose ends. And we might even extend that to comparing the number of loose ends in competing explanations. But I would resist that move because it is not at all obvious that fewer loose ends mean better explanation. It could mean a more limited explanation, which would defeat the objective of putting it all together.

In his new book (at this writing, in progress, material supplied by Ron Laymon in private communication), John Norton mounts a serious attack on the notion of inference to the best explanation. His primary moves are two: (1) there is no coherent account of explanation, and (2) this isn't what happens in science; that is, in choosing between theories, the choice is not based on an inference to the best explanation. He provides serious historical backup for 2. But I am not using inference to the best explanation as a suggested means for choosing between theories. I suggest we look for the most coherent story that provides the best explanation for how it all hangs together. This takes us back to Norton's first objection: there is no systematic general account of explanation. That may be true. His reason is that there are various forms of explanation in the sciences. That most probably is true. After all, each science has a different subject matter, employs different technologies, uses different theories, and so on. Why would you expect that we could come up with one account of explanation to handle this variety? His objection seems to be grounded in the Perennial philosopher's search for a single universal answer to a universal question. But if there is no such thing as science *per se*, then we should not expect there to be one universal account of scientific explanation.

Returning to problematics as historical entities, it is not enough merely to refer to "history," for there is the very matter of history itself to consider. To begin with, we use the word "history" in at least three different senses. Larry Laudan, in *Progress and Its Problems*, introduced a distinction between history of science (HOS) 1 and 2—HOS 1 is "the actual past of science" and HOS 2 is "the writings of historians about that past" (Laudan, 1977, p. 158). To draw this kind of distinction is a good idea, but it is not successful as phrased, since we cannot isolate the history of science *per se*. The history of science is embedded in a complicated past and cannot as such be separated from a lot of other things that were going on at the same time, for example, religious persecution. So, we must try a slightly different approach.

The PAST is what happened, the whole thing, all of it, every minute, second, and detail in the nonstop flow of time. History is the story we tell when we decide to select out certain items from The Past and credit them with some sort of importance.[19] It is like stopping a movie and snipping some frames and then splicing them together and trying to make that make sense.[20] Finally, there is another important ingredient to be considered and that is historiography.[21] Understanding the historiography an historian employs to write a history helps explain why it is this history rather than that one that is told.

In any telling of an historical story, the written history will be arbitrary to the extent the presentation of the facts and other factors deemed by the historian to be relevant will be a function of the historiography employed by the historian. Further, there is no rational selection procedure to pick an historiography. Why people choose to emphasize what they do in telling a story is not something that is the subject of rational discussion. When we challenge a tale woven by an historian we do so for a variety of reasons. Maybe we have discovered new "facts" that need to be included in the story. Maybe we prefer to emphasize different factors. Thus, for example, a story told by someone interested in the psychology of the actors will be a different story from the one told by someone interested in patronage. Further, History, that is, the story, will change as we uncover new facts we can show to be relevant and as we import new considerations into the story to make it hang together better with our understanding of what happened before and after. So, to the extent that we can never tell the whole story of what actually happened, and to the extent that the story we tell will be influenced by the selection of facts, the uncovering of new data, the realization that a different perspective needs to be added, History will constantly be changing as we attempt to tell the best, that is, most coherent, story possible. In the end, the best story possible is the one that hangs together best with what else is known—it is a matter of coherence in two ways, internal and external. But achieving coherence is difficult and, to a certain degree, arbitrary as well. A complete History will be a second-by-second telling of what happened everywhere at the same time and what happened before and after. That is, what was happening in China when Galileo was making his observations—not just in the emperor's palace, but in every hut and on every river, and this goes for everything in the universe. Such a story will be complete but unmanageable—just consider what it took for James Joyce to tell the partial story of what happened one day to three people in 1906 Dublin (Joyce 1922). In the end, the historian chooses what he or she needs from the data and attempts to weave a compelling tale. What he or she needs is a partial function of his or her historiography, conscious or not, as well as their motives; for example, Hume had a clear motive in writing his History of England, to ingratiate himself to the monarchy.

If it seems reasonable to reject the perennial philosophy, we open ourselves up to reviewing our relationships with our technologies in multiple ways. Just as our knowledge improves as our science does and our science improves as its technological infrastructures become more sophisticated, so too do our other institutions that face change as our technologies make earlier arrangements obsolete. One tendentious institution that may need revisiting is the Electoral College. One of the original justifications for the College was the need to overcome the geographical obstacles faced by the size of the country. Another was the desire to preserve states' rights. This was to be, after all, a *federal* system. The Electoral College was a means by which each state could contribute to the election of the president while ensuring that the people's voice was heard. Today our federal government tends to overwhelm individual states by the need to provide country-wide connections, national defense, and so on. So, the argument goes, given the development of rather sophisticated technologies, it seems both possible and reasonable to elect the president by a direct vote of the people. Whoever gets the most votes wins. We really don't need this cumbersome system that gives all the Electoral College votes of the citizens of a state to whoever got the majority in that state, essentially disenfranchising the losers. A direct vote, so the argument goes, is now possible and seems fairer. But, an objection goes, the Electoral College is part of our history, getting rid of it will amount to erasing part of our history. Well, comes the response, horses were a part of our transportation and farming history for a very long time, but having them replaced by tractors and automobiles and buses and trains hasn't deprived us of our history. New technologies make older ones sometimes obsolete, but it doesn't wipe history, so it goes.

The tales that follow in this story are selected because there is a point to be made against earlier tales of the development of science: those tend to leave out the role of the technologies scientists use and the technologists who invent and make them. For example, in his otherwise-first-rate analysis of discovery in the history of astronomy, Steven J. Dick (1973) notes, "In 1932 Rupert Wildt identified ammonia and methane in the spectrum of Jupiter." But we are not told how he did that, what instruments or techniques were employed, and so on. In the long run the case to be made is that we have the sciences we do because they are embedded in certain technological infrastructures that facilitate certain ways of doing things and discourage others. This leads me to the conclusion that *it is the technology that drives the science and not the other way around.* Or to coin a phrase—it is technology all the way down.

In the chapters that follow I will be exploring what constitutes a technological infrastructure, how new technologies affect what we mean by "see," how our technological infrastructures fail us, and how to reconceive the history and philosophy of science and technology. I will also extend the role of technological infrastructures beyond science and argue for the somewhat-radical claim that we are all immersed in a variety of technological infrastructures and that they are what make modern life possible, at least for a little while longer. I end with a look at the extent to which climate change is a product of an infrastructure that may be too big and complicated to be changed, leaving us with no options. Not only is such a reconceived history intellectually invigorating, but playing with the toys is fun!

NOTES

1. I understand that philosophy is often conceived of as a normative enterprise, but that does not mean we should ignore the real world.

2. As defined in my Thinking About Technology (2000), it is "humanity at work." It is a response to those who talk about Technology as if it were a thing in itself. Hopefully, my definition is so broad that it forces us to give up talking about Technology with a capital T and instead to look at specific technologies.

3. I will not address that issue. There is plenty of extant work on that problem, even to the extent of contesting the view that theories can be falsified, especially in the context of explicating or attempting to undermine Popper's version of falsificationism (Popper: 1959, p. 9).

4. My thanks to Nicholas Rescher for reminding me of Leibniz's *lingua universalis*.

5. While I do not think of myself as a Sellarsian in the sense of a follower, I have been heavily influenced in how I parse philosophical issues by his work. So, I often retreat to his writings to anchor some of my own thoughts. This, however, does not mean I am committed to parroting all of his views. I especially disagree with him on questions of scientific realism. See Pitt (2000, pp. 135–36).

6. I am employing a pragmatist theory of meaning here—where the meaning of a word or phrase consists in the total set of inferences one can make from a consideration of its use in a sentence. Thus part of the meaning of the sentence "The ball is red" is that the ball is not green, nor blue, and that what is red is not a bicycle, but a ball.

7. My thanks to Lydia Patton for locating the quote that eluded me for many moons.

8. Initially in "Philosophy and the Scientific Image of Man" (Sellars 1963, Chapter 1), but it is a theme that he continued to pursue or that motivated him to the end of his career.

9. Granted, many philosophers will interpret this claim as a charge that they are all Platonists. That is not what I am saying. Rather, it is this Platonic assumption that seems to have colored the course of philosophical inquiry. If not, then what is the justification for saying something like philosophy seeks general answers to universal questions?

10. For convenience I will sometimes refer to "science" as a shorthand for "the sciences." But in doing so I am not endorsing the idea that there is something out there called "Science."

11. Unfortunately, while Kuhn has remained popular for a variety of reasons, Hanson has largely fallen out of view.

12. My views on case studies and problematics have developed over several decades. Some of what follows can be seen again in Pitt (1992, 2001, 2007).

13. In a weak sense this view owes something to Imre Lakatos's notion of a research program without its systematizing aspects.

14. A lot of this comes from Pitt (2007, p. 272–289).

15. Concentrating on individuals is not a bad thing *per se*, as I will argue later. Rather, what is wrong is trying to show who influences whom. That game is not productive in revealing how the science was done.

16. It is also the case that when philosophers cite other philosophers, especially today, it is usually to disagree with them.

17. I refrain from speculation on why certain individuals are selected.

18. Thank you, Ron Laymon.

19. I thank Barbara Reeves for helping me articulate this difference between history and the past.

20. I realize that I am dating myself here—but there was a time when movies were made using something called "film" that was an extended connected length of plastic squares specially treated to hold an image.

21. Michael Seltzer forced me to consider this point.

Chapter 2

On the Origins of Scientific Observation[1]

In this chapter my aim is to begin an examination of the importance of optical instruments such as the telescope and the microscope on the emergence of our contemporary concepts of observation and on our understanding of the development of science, although I deal with the latter topic here only minimally. My underlying assumptions are two: first, already stated in Chapter 1, that the meaning of fundamental epistemic concepts such as "knowledge" and "observation" change over time, and, second, that understanding under what circumstances these concepts change is essential to understanding how science changes. I argue that a difference between "seeing" and "observation" emerged in the time period marked at one end by the development of the mathematical theory of perspective in the 15th century, and, at the other end, by the acceptance of the telescope (and possibly the microscope) as scientific instruments in the 17th century. Further, I argue that understanding the historical framework from which that difference emerged is essential to our appreciation of the manner in which science has developed and of our conception of scientific knowledge. I will also suggest, if not argue here, that fundamental epistemic concepts such as "observation" change primarily in response to non-theoretical developments, in this case because of the introduction of two crucial tools, the geometrical theory of perspective as a tool for rendering paintings more realistic and the telescope. The means for making this case, therefore, involve an examination of some features of Early Renaissance art, and a consideration of how it set the controls for seeing nature. In so doing I am attempting to extend the argument just offered in Chapter 1 regarding the complex infrastructure the doing of science requires. Again, contemporary science is fully immersed in such a set of infrastructures, but that does not mean that 17th-century science did not have its own infrastructure. What I am suggesting is that the theory of perspective and the telescope constituted part of that infrastructure.

The Copernican Revolution marks the beginnings of modern science. (See my *Galileo and the Book of Nature; Method Replaces Metaphysics*, 1992b.) It began with the publication of Copernicus's *De Revolutionibus Orbis* in 1543 and more or less ended with the publication of Newton's *Principia Mathematica* in 1687 and is marked by the mathematization of scientific inquiry. But it did not occur in isolation. A great deal of the cultural life of Europe was in turmoil during that time, as it continues to be. The transformations in the world of art occurred simultaneously with those in the world of science. If I am right that it is impossible to isolate specialized human activities from one another, there is no reason not to suspect that there were some influences going back and forth.

There is no doubt that the artistic representation of nature has changed over time. This is due in part to at least three different factors. (1) Changing senses of the term "nature." I have in mind here the difference between, for example, formal still lives on the one hand and, on the other, the opulent portrayals of lush New World forests. Let us call these styles of rendering nature. (2) A second factor contributing to changes in the artistic representation of nature is changing political agendas and religious influences on cultural values. Here one can talk about the revealing of such values as manifest destiny and the domination of nature in 19th-century American paintings. (3) Finally, there are the changing and evolving techniques for revealing and representing nature, from the development of the techniques of perspective to the use of the microscope and the telescope. Much has been written about (1) and (2) so I will not dwell on those topics (see for instance. . . .). Instead, I want to discuss some of the problems for representation that accompany the third factor noted earlier, the development of new techniques for investigating and for representing nature.

The introduction of the telescope and the microscope are part of the continuing process of evolving and perfecting the means of conveying one person's perception of nature to the public domain. The process goes as far back as recorded history. Furthermore, the appearance of the telescope and the microscope on the stage of inquiry is embedded in a long-playing activity marked clearly by the development of the theory and practice of perspective in the 15th and 16th centuries. This is a view I have argued for on several occasions (Pitt, 1992 p. 99–113, 1995). It has also more recently been put forth by Joseph Daube:

> This essay argues that the advent of linear perspective, ca. 1425, when Filippo Brunelleschi painted a small panel of the Florentine Baptistery by applying the geometric rules of optical mirror reflection, was more than just an artistic event.

> Indeed, it subsequently had the most profound—and quite unanticipated—influence on the rise of modern science. Surely, by 1609, Galileo would not have understood what he saw when observing the moon through his newly invented optical telescope, then called the "perspective tube," had it not been for his training in perspective drawing. (Daube 2015)[2]

Further, it should be noted that the development of the mathematical theory of perspective in the 15th century marks the emergence of mathematical physics in the 17th century as very much a Johnny-come-lately in the project of capturing mathematically the secrets of the Book of Nature. More to the point, I argue later that mathematical physics flowered in the 17th century, in large part because of the prior development of the geometric theory of perspective.

Turning now to the particular concern of this chapter, my question is this: *what is the relation between the development of the scientific concept of observation and the introduction of the telescope and the microscope in the 17th century*? In other words, did people see nature differently after the introduction of these instruments? There are several ways to attack this problem. First, we could attempt to determine if the depiction of nature in art changed in ways we can attribute to the use of such instruments. Second, we could attempt to find written evidence that the produce of these instruments presents problems for investigators of the period. Third, we could look for evidence that shows that the universe actually was being perceived differently. I think I can show that the proper response to the first question, which asked if the art of the period revealed any differences, is no. The relevant changes in artistic style occurred earlier, in the 15th century, following the development of the theory of perspective, long before the invention of the telescope. I can also show that there is a significant problem posed for the observer using the instruments. Finally, there is evidence that the instruments allowed their users to see nature in different ways, at least where the telescope is concerned. Let's consider each of these in turn.

Why look at the art of the period to see if these instruments made any kind of difference in seeing? Renaissance art, with its highly developed theoretical counterpoint, demonstrates an extremely sophisticated sense of seeing. It is also a deeply theoretical activity. There are numerous treatises on painting and perspective by, to select a very few, Leon Battista Alberti, Gregorio Comanini, Giovanni Baptista Fonteo, Leonardo Da Vinci, Albrecht Durer, and Carlo Ubaldo (for critical historical accounts of the relationship between geometry, art, and perspective, see Ivins [1946] and Pedoe [1976]). And, as noted earlier, much of the most significant theoretical work was completed before the invention of the instruments I have identified. The level of representational accuracy found in such places as illustrated texts as well as paintings and frescos cannot be disputed. Furthermore, none of these items show any influence of the telescope or the microscope on their techniques or on what was represented. Later we do see representations of scientific instruments in paintings, but that by itself shows no change in their conception of observation *per se*.

On the other hand, I would like to suggest that the conceptual shift in physics from the qualitative Aristotelian approach to a mathematical physics in the 17th century was facilitated by the prior 15th-century geometrization of art through the introduction of the geometric theory of perspective. A stronger thesis holds that the development of a scientific conception of observation conceptually required the prior development of this mathematical dimension in art. Let me elaborate on each of these suggestions.

Elsewhere (Pitt 1992b), I have argued at length that the success of Galileo's arguments in his *Dialogue on the Two Chief World Systems* depended to a large extent on his appeal to geometric proofs because geometry was perceived as a neutral tool by protagonists on both sides of the debates. But how did this situation regarding the general acceptance of geometry come to pass? Aside from providing an established sense of proof, geometry also had already established its credibility in art and in music theory, that is, in the cultural dimensions of life. Vincenzo Galilei, Galileo's father, was a well-known lutenist and music theorist who established the ratio of notes to chords mathematically and experimentally. We know that during the Renaissance few intellectuals compartmentalized their interests and/or specialized in the manner we do today. All the various features of life were open to investigation and integration so that the importation of techniques from one type of activity to another different type of activity was a natural mode of operating. This leads to the second suggestion discussed earlier, that the development of the geometric theory of perspective was conceptually necessary for the development of a concept of scientific observation.

First, consider what we require today of a scientific observation. It is a deliberate attempt to determine specific characteristics of an object or a situation characterized by the attempt to convey those features to a public forum. A scientific observation that cannot be relayed to others is not an observation. Now consider what Leon Battista Alberti, the first to publish a full-scale geometric account of the principles of perspective, had to say on the subject of constructing the vanishing point in a painting, the central and most distant place in the image to be constructed.

> First of all, on the surface on which I am going to paint, I draw a rectangle of whatever size I want, which I regard as an open window through which the subject to be painted is seen; and I decide how large I wish the human figures in the painting to be. I divide the height of this man into three parts, which will be proportional to the measure commonly called a braccio [arm]. With this

> measure I divide the bottom line of my rectangle into as many parts as it will hold. . . . Then I establish a point in the rectangle wherever I wish; and [as] it occupies the place where the centric ray strikes, I shall call this the centric point. The suitable position for this centric point is no higher from the base line than the height of the man to be represented in the painting, for in this way both the viewers and the objects in the painting will seem to be on the same plane. Having placed the centric point, I draw lines from it to each of the divisions on the base line. (Quoted in Egerton [1976], pp. 42–43) see Figure 2.1

We find Alberti first defining the space to be seen by using what we call today the technique of framing. Then by establishing the vanishing point within the framed space we are told how he intends to provide for objects to be seen, nearer and farther. The crucial point is the framing, which then permits the vanishing point to be established. Essentially, he is constructing not

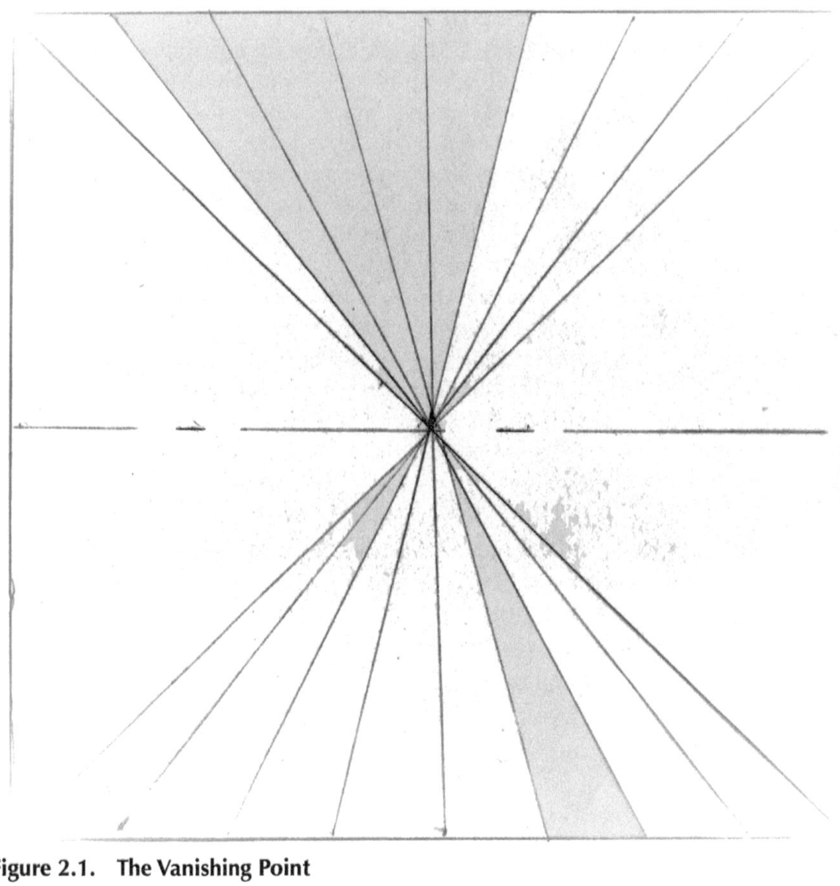

Figure 2.1. The Vanishing Point

just the space, but also the way in which we will see what is in the space. He is defining what will be seen and how it will be seen. These, at least to me, are the conceptual first steps of formal scientific observation. Developing the relational theory of how to do this is a necessary condition for establishing the epistemic significance of observation.

Let us now return to the second of the three questions raised earlier about the impact of the telescope and the microscope. The first question concerned the impact on the artwork of the period. The answer there was negative, but I suggested instead that the theory of art as augmented by the geometrical theory of perspective instead had consequences for the concept of observation, as will the telescope and the microscope.

The second question was, Is there evidence that the results of using these instruments provided problems for the observers? Yes, there is. Let me quote from Robert Hooke's preface to his *Micrographia*, the first full-length treatise on microscopy, published in 1665.

> What each of the delineated Subjects are, the following descriptions annext to each will inform, of which I shall here, only once for all add, That in divers of them the Gravers have pretty well follows my directions and [draughts]; and that in making of them I have endeavored (as far as I was able) first to discover the true appearance, and next to make a plain representation of it. This I mention the rather, because of these kinds of Objects there is much more difficulty to discover the true shape, then of those visible to the naked eye, the same Object seeming quite differing, in one position to the Light, from what it really is and may be discov'd in another. And therefore, I never began to make any draught before by many examinations in several lights, and in several positions to those lights, I had discover'd the true form. For it is exceeding difficult in some Objects, to distinguish between a prominency and a depression, between a shadow and a black stain, or a reflection and a whiteness in the colour. Besides, the transparency of most Objects renders them yet much more difficult then it they were opacous. (Hooke 1665)

The problem for Hooke is twofold. Not only is there genuine difficulty determining what is being seen, but there is the additional problem of communicating what one has seen to an audience. There is a third problem not confronted by Hooke, but it plagued Leeuwenhoek. Hooke could at least draw. From his detailed sketches he could coach his engravers. Leeuwenhoek could not draw. Instead, he relied even more heavily than Hooke on the services of an illustrator, in this case, one Willem van der Wilt, to convey the images of what he saw to the Royal Society (Dobell, 1932 p. 342).[3] This problem of the observer not being able to draw introduces one more level of indeterminacy to the problem of communicating what was seen.

Neither of these difficulties was resolved until the 19th century with the creation of cell theory in 1838 and then refined staining techniques later in the century. There is no real explanation for the delay in the development of staining techniques. The problem was the need to stain certain features in the slide and not others.

It was Joseph von Gerlach who in 1858, while experimenting with solutions of carmine and leaving a section of brain tissue in a dilute carmine solution overnight, reported good differential staining of the nucleus and nuclear granules compared with little or no staining of the cytoplasm and intercellular substance. He concluded that previous staining techniques had been too concentrated and also noted that the dye was absorbed by specific cellular elements and could not be washed out (Heinrichs 2009, p 26).

With the telescope the problem is slightly different, depending on the object being observed. In *Siderius Nuncius*, Galileo goes to great length to explain to his audience how he arrived at his conclusion that the moon has a rough surface. He showed them nothing new. What he did was demonstrate how to interpret what everyone saw, namely the shadows of mountains and craters. He did this by appeal to what was familiar to all, namely our own experience with shadows of mountains on the earth. But prior to employing his analogies, he cautions his readers to make sure they have the right type of instrument, less they not be able to replicate his observations. And so he says,

> In this matter, it behooves all those who wish to make such observations to be forewarned. For it is necessary first that they prepare a most accurate glass that shows objects brightly, distinctly, and not veiled by any obscurity, and second that it multiply them at least four hundred times and show them twenty times closer. (Galileo 1610)

Here I see Galileo following up on Alberti's program of framing the observation, this time by requiring that the proper instruments be obtained and used in specific ways. It is a natural step—one that seems intuitive to us. We have been raised with the idea of the replicability of scientific observations. But here we see the first steps being taken that establish the conditions for achieving these results.

Gerard Turner (1980) argues that the problem of representing accurately what the observer saw through the microscope was solved with the development of photography. I am not convinced, since this suggests that the problem is merely one of being able to render accurately what one sees. But as anyone who has ever looked through a microscope, or for that matter a telescope, knows, you have to be taught how to see through these instruments. It is my sense that the problem was more than being able to convey to an engraver what was seen by the observer. What the observer saw was itself a problem. Thus, Leeuwenhoek, in a letter to Oldenburg complaining he can't draw, continues,

> Yet I am to blame, because I can't draw; and secondly, because I am resolved not to let anybody know the method I use for this purpose and so I just make only rough and simple sketches with lines, mostly in order to assist my memory, so that when I see them I get a general idea of their shape; besides, some of the forms I see are so fine and small, that I don't know how even a good draughtsman could trace them, unless he makes them bigger. (Quoted in Dobell 1932, p. 67)

Previously I listed three questions concerning how the microscope and the telescope affected the notion of observation. My final question is, "was there any evidence to support the claim that the introduction of these instruments actually made a difference in how their users came to see nature?" The answer is yes, but we need to divide the question and look at each instrument separately. In the case of the microscope, it seems clear that there emerged a fundamental problem of seeing something that one could not then simply show to another. Even if the other observer were present, it would be very difficult to determine if he saw what Leeuwenhoek saw, for example. Seeing through a microscope then becomes something private, whereas observation as captured in artistic representations of nature hitherto had been a reasonably public affair. There is a certain oddity in saying that you are seeing something under the microscope, but that you can't precisely communicate what it is. The realism of Renaissance art, emphasizing the theory of perspective, for example, for the purpose of training visual artists to communicate accurately the nature they saw in front of them, existed in a somewhat-different world from that of the observing scientist. The artist can always check to see if what he or she is putting on canvas is an accurate rendering of a scene by asking another observer to look. But the scientist working through a telescope is not in the same position.

In this sense, the new instruments of scientific observation created a strange paradox. Scientific knowledge is supposedly public knowledge, knowledge that is replicable. Furthermore, it is based on experience, where observation is the paradigmatic form of experience. And yet, with the introduction of these instruments, which extended the observational power and base of science, we see its public dimension weakened, if not undermined. The problem is exacerbated when we come to the electron microscope (see Chapter 5).

Similar problems afflict the telescope. It is not just a matter of accurately reporting what one has seen. Interpretation is necessary. Galileo, for example, knew that the most successful way to argue for his interpretations of his observations of the heavens was to draw analogies with earthly phenomena.

But when the analogy broke down, the situation became complicated. Consider the problem of explaining the extra stars in a telescopic observation of the Pleiades, as opposed to naked eye observation.

But there was yet another aspect of seeing nature that the telescope introduced, which in the 17th century fundamentally altered the very nature then seen. The images Galileo reported were tantalizing. It was not long before there was demand for better instruments to produce better images. The result was the replacement by the middle of the 17th century of the Galilean telescope with the astronomical telescope. The difference between the two was that the astronomical telescope had a convex ocular rather than a concave one. Not only did this give the astronomical telescope a larger field of view, but as Van Helden (1989) points out, since it was also a larger lens, it was now possible to introduce a micrometer into the instrument. This then made it possible to measure distances between planets and among stars. One could no longer look at the heavens in the same way. One could argue that ancient observers using the Aristotelian model of the heavens also had a three-dimensional universe—but theirs was very minimally populated with only the earth, the planets, and the sphere of fixed stars. Once the sphere of fixed stars is replaced by a three-dimensional relation among the stars themselves, the whole picture changes, gaining depth and complexity.

The possibility of measuring the universe is a conceptual step necessary for the later introduction of the concept of an infinitely expanding universe. Not only does it change our conception of the universe but also our understanding of what it is possible to see—for example, three-dimensional relations among the stars—as opposed to the stars being fixed on a rotating sphere with only left-right-up-down relations. When I first got my telescope I trained it on the Pleiades—but I could barely find the seven sisters among the 100 million stars that popped into view. The universe becomes a different place when viewed through a telescope. Thus, while these marvelous instruments made it possible to see nature in more detail and to transform our understanding of the nature we see, infinitely large and infinitely small, they also, again somewhat paradoxically, introduced us to the idea that there were limits to what we could see, limits imposed by the instruments—and by our understanding by virtue of lack of appropriate theories to help interpret the images.

So far, I have tried to emphasize the manner in which two instruments, the microscope and the telescope, forced their early users to acknowledge the fact that seeing nature through them was not merely a matter of faithfully rendering what was seen as was done in a drawing. With the introduction in the 17th century of instruments to aid seeing, interpretation now becomes a crucial part of the game and it also becomes a crucial part of our understanding of observation as opposed to mere seeing. The artists of the Renaissance worked on the science of seeing nature as accurately as possible—that was the motivation for the development of the mathematical theory of perspective. The scientist of the 17th century started on the long road to specifying what it meant to observe nature, which turned out to be a complicated job of balancing seeing, interpreting, instruments, and measurement. In order to undertake this job, they had to rely on what the artists had earlier achieved in terms of mathematizing visualization.

That being the case, I urge the strong thesis that scientific observation is impossible without doing something, that is, framing, measuring, and/or interpreting nature. A scientific observation is not a mere seeing. Rather it is a complicated and active process by which the observer acts on nature in order to make it comprehensible. And it is made possible through technological innovation. Seeing nature, on the other hand, is what artists and naturalists do. It is no less valuable. But, in the long run, it is different because, while conceptually and physically necessary for observation, it is limited to seeing.

To appreciate the full force of the strong thesis we need to move forward several centuries to the invention of the electron microscope—where we still talk about seeing things—but what we are really doing is quite different from looking through a telescope or using a handheld microscope. But first, we will discuss the effect of Galileo's telescope on the conceptual scheme of his times.

NOTES

1. Parts of this chapter appeared as Pitt (2016b).
2. I thank Ron Laymon for bringing this quote to my attention.
3. I leave it to the reader to determine whether or not the illustrator *qua* illustrator counts as a technology.

Chapter 3

Galileo and the Telescope

A paradigmatic example of technological innovation effecting scientific change remains Galileo's use of the telescope to investigate the heavens. What follows is *not* a case study. It is a history of a sequence of events.

In 1610 Galileo published *Siderius Nuntius* (*The Starry Messenger*), in which he reported his telescopic observations of, among other things, the moon and Jupiter. There is consensus now that Galileo did not invent the telescope. Who invented it is a question that has not been definitively answered either, but Sir Harold Spencer Jones, astronomer royal in his foreword to Henry C. King's thorough *The History of the Telescope* (1955), credits Giambattista della Porta with building the first telescope *circa* 1590. Hans Lipperhey, a Dutch spectacle-maker, had applied for a patent, not granted, for it in 1608, and he is sometimes claimed to be its inventor.

The importance of the discoveries Galileo detailed cannot be underestimated. There is much discussion as to whether he can be said to have caused the Aristotelian worldview to be abandoned. That is the sort of question that cannot be answered since there were other fires burning, for example, the Reformation and Counter-Reformation. There is, however, no doubt that his work contributed greatly to the change from a geocentric worldview to a heliocentric worldview and that the telescope was critical to that shift.

How Galileo came to build that instrument is also subject to discussion. The popular account is that he was told about the instrument from an itinerant peddler and using his incredible mental powers figured out how it worked all by himself. Galileo's own account, as printed in Stillman Drake's *Galileo at Work*, somewhat supports this account. In June 1609, Galileo was in Venice.

> In Venice, where I happened to be at the time, news arrived that a Fleming had presented to Count Maurice [of Nassau] a glass by means of which distant

> objects could be seen as distinctly as if they were nearby. That was all. Upon hearing this news I returned to Padua, where I then resided, and set myself to thinking about the problem. The first night after my return I solved it, and on the following day I constructed the instrument and sent word of this to these same friends at Venice with whom I had discussed the matter the day before. Immediately afterward I applied myself to the construction of another and better one, which six days later I took to Venice, where it was seen with great admiration by nearly all the principal gentlemen of that Republic for more than a month on end, to my considerable fatigue. (pp. 137–38)

Drake places the beginning of Galileo's observation of the moon at the end of November 1609. But Galileo was not the only one using a telescope. Kepler received a telescope in 1610 and started work on his optics in order to refine it. Scheiner was also using one, as was Francesco Fontana, a Neopolitan. Telescopes were everywhere, and they were being used to observe sunspots, the moon, the planets, and so on. That being the case why did the publication of Galileo's results cause such a stir?

At this time, the prevailing worldview placed the earth at the center of the universe with the moon and the rest of the planets rotating around it. The stars were fixed on a sphere that rotated around them all. While Copernicus had published his geocentric theory of the structure of the universe in 1546, it was not well received, defying the established view and challenging scripture. Further, he had no physics to explain how the planets could move around the sun. His account was, as best, a complicated mathematical theory. And his student, Rheticus, wrote a preface in which he described the theory as a purely mathematical account, which didn't help matters.

In 1613, Galileo published his three letters on sunspots in response to the attacks launched on the possibility of sunspots and on the Copernican system by Scheiner along with Scheiner's three letters. Then, in 1616, Galileo went to Rome on his own accord, which was a mistake.

In the preceding years, a minor controversy had developed stemming from a letter Galileo wrote to Castelli in which he tried to reconcile scripture with Copernicus's heliocentric view.[1] As was the practice at the time, the letter was copied and circulated. An incomplete copy of the letter found its way to Rome and caused some concern. Even when a copy of the complete letter was sent, it did not settle the matter. Cardinal Bellarmine (later Pope Urban VI), a supporter of Galileo, warned Galileo to speak only as a mathematician. While in Rome, there were a series of incidents that resulted in Galileo being told by Bellarmine not to defend Copernicus and that if he insisted on defending Copernican views he would be arrested.

It was quite clear that, while Galileo had friends in high places, there were also serious concerns that the Copernican view threatened crucial tenets of Catholicism, primarily that the earth was at the center of the universe. Galileo's revelations about the moon and the Medicean planets suggested a defense of Copernicus, but more importantly they challenged the philosophical basis of geo-centrism. The heavens were thought to be perfect, while change characterized the earth. Growth and decay was the way of all features of our planet. The planets and stars circled the earth because moving in a circle was perfect motion. On the theological side, man is God's greatest creation and God placed him at the center of the universe as evidence of man's importance. If the earth is just another planet circling the sun, man's significance is diminished, and his relation to God becomes uncertain.

In effect, the observations produced by the telescope forced a rethinking of an entire worldview. Clearly, this is an extreme example of technological innovation forcing scientific change. We have not seen another example of such magnitude since, although it could be argued that the development and use of artificial intelligence and machine learning is coming close. But it illustrates clearly one of the central theses of this chapter.

Another thesis proposed here is that the doing of science is made possible by the technological infrastructure in which it takes place. Exploring what this amounts to will then put us in a position to look at some of the issues and events charactering the sciences. In what follows, I will examine some aspects of what this amounts to.

NOTE

1. This was a precursor to the more extended defense of the compatibility of science and scripture known as *Letter to the Grand Duchess Christina.*

Chapter 4

The Technological Infrastructure of Science

We all know the old story: scientists do basic research and technologists apply it. I have always been troubled by this account because something about it doesn't ring true. In particular, I have been curious about how the move from basic science to applications takes place. The results of basic scientific research are published (when they can be published) in very specialized venues using very specialized languages not readily accessible to most mere mortals. I am not suggesting that engineers can't read this literature, but simply asking if, given their other responsibilities, do they take the time to do so?[1] It is not clear to me that having made some discovery or other the scientist then sends out an e-mail and tells his or her engineering colleagues "now you can do this or that."[2] Nor is it clear that engineers keep close track of the burgeoning scientific literature to find out what's new and have immediate "ah, ha!" moments, or, even later, "duh" moments. There is also the complex problem of intellectual property rights, finding interested investors, manufacturers, distribution routes, and others. So, when push comes to shove, I think the old story is not only a false story, but highly misleading.

I offer a different story: if the technological infrastructure of science is, in part, the product of engineering research and hands-on design and inspection, and the research that makes it possible to build both labs and instruments comes out of engineering research, then engineering research is more fundamental than scientific research. But there is more, maybe this way of conceptualizing the issue is wrong from the start.

Although he hasn't said this explicitly, the argument I am proposing is congenial to the views Peter Galison develops in *Image and Logic* (Galison 1997). In particular, I have in mind his distinction between the inner and outer lab. The outer lab I take to be amenable to the notion I introduced in my *Thinking About Technology* (2000), the technological infrastructure of science. What the technological infrastructure idea is supposed to capture is the range of things that make science possible: funding agencies, universities, private corporations, technicians, labs, graduate students, journals, and so on. I will proceed by returning to my motivating issue here, which is whether or not it is correct to think of the relationship between science and engineering as one of subservience. I am going to begin by looking at a couple of examples and argue for a more encompassing view.

Some cases are easier to understand than others—that is, some discoveries more readily suggest applications than others. This can happen in many ways, but looking at two examples will yield the general idea. In the first case, a scientist can be looking to achieve a specific end, which itself is an application. Thus, consider a microbiologist who set out to construct a bug that would eat oil. He started on this after hearing of a particular disruptive oil spill when a tanker went aground. After successfully creating the oil-eating bug, he set up his own company, rented some space at the Virginia Tech Corporate Research Center, hired some graduate students to make the things, contacted oil companies to inform them of the product, and started selling it. What is missing from this picture is an engineer. Moreover, while a close examination of the process employed to create the bug deeply resembles a classic engineering design process, complete with feedback loops, our actor is a biologist, not an engineer. In this case, the line between scientist and engineer is clearly blurred, given the standard story. But the positive result of this story is that it opens up the possibility that in some cases, the so-called scientific method is more like an engineering design process than some idealized and false view of how scientists do their work.

A second example of how scientists connect to applications is the "oops!" case. This occurs when a mistake is made or an accident occurs in a lab and an unintended result comes up that has immediate applications because of the result itself. The process by which this discovery makes its way into the public domain may or may not involve engineers down the road, but the awareness of its applicability does not. For example, consider the case of penicillin.

> As the story goes, Dr. Alexander Fleming, the bacteriologist on duty at St. Mary's Hospital, returned from a summer vacation in Scotland to find a messy lab bench and a good deal more.
>
> Upon examining some colonies of Staphylococcus aureus, Dr. Fleming noted that a mold called Penicillium notatum had contaminated his Petri dishes. After carefully placing the dishes under his microscope, he was amazed to find that the mold prevented the normal growth of the staphylococci.
>
> It took Fleming a few more weeks to grow enough of the persnickety mold so that he was able to confirm his findings. His conclusions turned out to be phenomenal: there was some factor in the Penicillium mold that not only inhibited

> the growth of the bacteria but, more important, might be harnessed to combat infectious diseases.
>
> As Dr. Fleming famously wrote about that red-letter date: "When I woke up just after dawn on September 28, 1928, I certainly didn't plan to revolutionize all medicine by discovering the world's first antibiotic, or bacteria killer. But I guess that was exactly what I did." (From PBS News Hour, September 27, 2013, by Dr. Howard Markel, 2013 Internet)

He called it penicillin. Unfortunately, naturally occurring penicillin isn't very stable and thus not very useful. Fleming had found a wonder drug but couldn't do much with it. Luckily just three years later two Oxford researchers created a stable form, and today it's one of our most important tools in the fight against disease.

Consider now an example of a discovery that was delayed in its application and why.

> A Post-it note is a small piece of paper with a strip of low-tack adhesive on the back that allows it to be temporarily attached to documents, walls, computer monitors, and just about anything else. The idea for the Post-it note was conceived in 1974 by Arthur Fry as a way of holding bookmarks in his hymnal while singing in the church choir. He was aware of an adhesive accidentally developed in 1968 by fellow 3M employee Spencer Silver. No application for the lightly sticky stuff was apparent until Fry's idea. The 3M company was initially skeptical about the product's profitability, but in 1980, the product was introduced around the world. Today, Post-it notes are sold in more than 100 countries. (science.howstuffworks.com)

The point of these two examples is to suggest that to understand the move from scientific discovery to practical application needs more than hand waving at science and technology as such. We have already observed that the process of going from a discovery to a practical application is more complicated than the standard story would lead us to believe. While complicated, it nevertheless seems possible to spell it out using the standard story. However, I want to argue that even doing so will still give us a skewed picture.

The picture is skewed because it starts with the scientist. It suggests that the scientist does research and comes up with discoveries, but it does not fill out the picture as to what is entailed by saying the scientist does research. In order for a scientist to conduct research, he or she generally needs a lab. It can be as simple as a computer in a room, or as complicated as a radio telescope, but to say a scientist conducts research entails that there is a context (sorry about that) in which that research is done, even field scientists who study the behavior of the great apes treat the environment in which the apes live as their lab. Once we open that door, the entire picture changes.

In *Thinking About Technology* (2000), I introduced the notion of the technological infrastructure of science. I characterized it as "an historically determined set of mutually supporting artifacts and structures that enable human activity and provide the means for its development" (p. 122). If we restrict our attention to science (because there are technological infrastructures at play in culture in general, more later), as noted earlier, parts of this complex are the labs, graduate students, technicians, instruments, universities, and funding agencies that make modern science possible.

Consider what is involved in hiring a new scientist at a typical American research university. I am not talking about the hiring process, that is, the means by which the individual hired is selected—but rather the rest of the process that must be completed before the offer is accepted: the support package offered to the new potential hire as an enticement to accept the offer. No active researcher would think of accepting a position without being guaranteed a lab, that is, a particular space, and startup money to equip the lab with the appropriate equipment needed to conduct his or her research, to hire a technician or two or three and to support at least a couple of graduate students. As of 2010, the typical "startup package" at my university for a new PhD in one of the sciences or in one of the areas of engineering, coming straight out of school and off a two-year postdoc, cost approximately $400,000. It gets way more expensive for senior researchers.

Now let us unpack this a bit further. Laboratory space is expensive. Depending on the research to be done, there will be a water supply and sinks, exhaust hoods, computers, isolation spaces, and others, all housed in buildings meeting more stringent building codes (meaning costing more to build) than your typical classroom building. The differential here just for the costs of the buildings is $324/square foot for a classroom building versus $621/square foot for a laboratory building (unequipped) (2019 figures). Doing science is expensive.

Second, part of the startup package involves the money needed to fund the research. But it is also money that provides the time for the researcher to develop a research program and to write grant proposals to support further research once the startup monies run out. That means there have to be sources for that funding. I would argue that the sources of funding, like National Science Foundation and National Institutes of Health (NIH) on the public side and various foundations and companies on the private side, also constitute part of the technological infrastructure of science. They are enabling systems. Moreover, by virtue of having the money and issuing calls for proposals in certain research areas on certain topics, they not only enable scientific research, but also to a large extent they control its direction. In the United States, at the time of this writing, federal sources of research funding may not fund embryonic stem cell research on strains of embryonic stem cells recently developed following a 1996 NIH policy. (See NIH Stem Cell Information page.) This is having an interesting effect in two directions. (1) It is forcing certain kinds of research to be suspended or terminated for lack of funds. (2) It has pushed individual states like California to appropriate the funds themselves for such research, thereby putting them in the position to attract researchers in these areas away from states where they cannot do their work, thereby making the universities and research centers in California a major force in this area of research. So, funding sources make a difference in how science is done and what kinds of scientific research will be done and where it will be done. The picture of scientific research is getting messy.

Let us return to the lab—for convenience sake, let's make it a university lab. The preceding picture is too simple—we don't just give the new researcher a lab and some money. The buildings have to be designed, built, and inspected to meet formal specifications. The instruments have to be designed and built. In short not only are the funding agencies needed, the engineers who translate architects' designs into buildings and who make sure they meet building codes, as well as the teams who design and oversee the building of instruments are essential infrastructure components for scientific research. The materials that are used in the buildings are the product of engineering research for the most part, and that research requires the same kind of support as scientific research does—labs, technicians, graduate students, funding agencies, and others.

To be a bit more specific, the spaces where scientific research is done do not simply appear out of nowhere. They are designed spaces. And then they are built spaces. I am deliberately making a distinction here between designing the space and building it. It actually needs to be a threefold distinction: designing the space, figuring out how to build it, and building it. Architects, if they figure into this process at all in a significant way, work in the first part, designing the space. For the most part, the most significant part of the work involves figuring out how to make the proposed design work—and that is an engineering job. Architects are notorious for drawing lines that appear to connect and leave it up to engineers and builders to figure out how to actually make them connect. It is of no small note that the most successful architectural firms today—what are called full-service firms—involve both architects and engineers in the process of getting a building from plan to fact; sometimes they also supply the builders. So, the very spaces in which science is conducted are heavily influenced by engineers. But there is more, for the materials used to build these spaces are constantly being improved thanks to engineering materials research. And that research is conducted in much the same way scientific research is—in specially designed spaces, and so the cycle spirals upward and beyond. Peter Galison's account of laboratory design in *Image and Logic* speaks directly to this point.

What engineers do and how they do it, to coin a phrase,[3] is fundamental to what scientists do and how they do it. So far I have only addressed the spaces where scientific research occurs; if you will, the building of the spaces. But if we will also look inside the science lab, we find the footprints of the engineers all over the place. Maybe not in labs of the gorilla researchers, but in the labs in the buildings we have been discussing we find instruments. Sometimes instruments are designed and built by scientists. If you will allow the anachronistic use of the term, if we consider Galileo a scientist, then we also have to contend with Galileo the instrument maker. One of the sources of income he relied on was the sale of instruments he not only invented, or made popular, but also built and sold, such as his military compass as well as his telescope. And it is well known that many contemporary scientists build their own experimental apparatus, pulling this and that off the shelf, which is one of the things that make the replication of experimental results so difficult; the shelves in different labs contain different materials.

Nevertheless, when it comes to buying equipment from commercial suppliers to equip your scientific lab, engineers are involved up to their elbows. For in the production of standardized laboratory equipment engineers play a major role, for these instruments are their provenance. In short, the scientist could not do his or her job without the engineer. There would be no appropriate space in which to work. The development of quality materials would be greatly delayed. If anything, there would be fewer and more poorly made instruments as well as what else is needed to fill out a functioning lab, instruments needed to conduct that work without engineers working independently.

Now I want to be clear about this. I am not saying that from the beginning of time, engineers were central to the doing of science. The thesis I am reaching for is this: *modern* science relies on this technological infrastructure, in which large components involve work in which engineers play a major role. It would be an interesting doctoral thesis to trace the historical development of the split of *media scientia* (middle sciences, something in-between theory and craft) into science and engineering. Something obviously happened in the 16th–17th centuries. The *media scientia* were already recognized as doing something applied—both Da Vinci and Galileo were often employed as what we would today call engineers—working on military fortifications while doing multiple other things, like painting and writing music. But to talk that way may be too simplistic. Why should we assume that there was a split into something like science and engineering from something like the *media scientia?* Maybe things don't happen that neatly if you are looking to draw straight lines, maybe lots of things are going on that make it look simple in retrospect, but it's not.

So, the bottom line here is that simple generalizations about the relation of this to that need a more nuanced historical analysis that goes deeply behind the surface to uncover what really is going on. I hope I have provided a schematic for the kinds of details that need to be examined—it is not presented as the full story by any means. There are a couple of problem areas here that we need to be sensitized to: (1) the reification of human activities—for example, science as somehow something that doesn't take place in a time and place; (2) Galison's idea of how science changes, not all at once, but different parts changing at their own pace, works here; (3) the politics of priority—I haven't talked about this, but it is worth raising, even in passing: as any sociology undergraduate major will tell you, there is a competition in society among groups for some kind of social recognition. In our story, it is alleged to be between science and engineering. But that just may be the wrong way to frame the discussion. It assumes there are these things that are called science and engineering, when in fact they are complexes of great complexity. I would much rather talk about the particulars, the people and what they do. The following argues that there is no such thing as science and no such thing as technology simpliciter.

NOTES

1. Actually, that is what graduate students are for—keeping their major professors up to date.

2. I am using engineers as my class of technologists here because they are readily identified with a variety of technologies. I recognize that there are problems with this approach and that engineers are just a convenient subgroup to consider.

3. *What Engineers Know and How They Know It* is the title of a groundbreaking work by Walter Vincenti (1993).

Chapter 5

"Seeing" at the Nano-Level

As we saw in chapter 2, while technological innovations often open up our vision and understanding of the world, they also expose limitations. These limitations constitute problems for the observer because they restrict what can be offered to the general public. A vivid example of this sort of problem is posed by the electron microscope.

Optical microscopes reached the limit of their usefulness for probing the realm of the very small in the late 19th century because they reached the limit of their ability to resolve an image.

> In general, a given type of radiation cannot be used to probe structural details much smaller than its own wavelength. This is a fundamental limitation of all microscopes. The ultimate limit to the resolution of the light microscope is therefore set by the wavelength of visible light, which ranges from 0.4 um (for violet) to .7um (for deep red). In practical terms, bacteria and mitochondria, which are about 500 nm (0.5 um) wide, are generally the smallest objects whose shape can be clearly discerned by the light microscope; details smaller than this are obscured by effects resulting from the wave nature of light. (Alberts et al. 2002)

The electron microscope was invented in 1931 by Max Knoll and Ernst Ruska at the Berlin Technische Hochschule. It was able to overcome the limitations of the optical light telescope by using electrons as the effective visibility medium. The world it opened up for us is the nano-world, the world of the very small.

> One nanometer is 1-billionth of a meter. To get an idea of the size, we can compare some familiar things. The diameter of an atom is about ¼ of 1 nanometer. The diameter of a human hair of 10,000 nanometers. The protein molecules,

> which are so important, so critical to life, are several nanometers in size. Moving to man-made things. The smallest devices on commercially available chips are about 200 nanometers, whereas the smallest experimental chips are approximately 10 nanometers in their smallest dimension. (Nanotechnology: The State of Nano-Science and Its Prospects for the Next Decade, p. 3)

The nano-world is currently brought to us through the technology known as the electron microscope. It differs from the optical or light microscope in many ways. But, for our purposes, perhaps the most important difference is that one does not look through an electron microscope. Actually, the "observer" does nothing. Today what is "observed" is displayed on a computer screen, suitably enhanced for the human "observer" to "see" it. There are different types of electron microscopes. The ones we will be concerned with are called scanning tunneling microscopes (STMs), invented in 1981 by Gerd Binnig and Heinrich Rohrer. According to Rasmussen and Hawkes,

> An electron beam that is small compared with the imaged areas passes over the specimen in a regular pattern, and a picture of specimen surface is reconstructed on a video tube. . . . interaction of the beam with the specimen produces varying intensities of backscattered and secondarily released electrons for each position in the scan, and these are registered by a detector placed appropriately near the specimen. . . . All electron microscopes depend on the capacity of the magnetic and electric fields to alter the path of electron beams according to the laws of optics. (in Bud and Warner 1998, p. 383)

The limitation in question here is that we cannot "see" what exists at the nano-level. We must rely on the STM to produce an image that we can in fact actually look at.

A couple of things should be noted here. First, there is the role of the unmentioned computer technology, hardware and software, that turns the activity of a beam of electrons into an image. Second, the language of observation changes meaning when we refer to STM-generated images as "images" and that we "see" the thing imaged. Let us look at these two issues separately.

The means by which a beam of electrons is converted into an image on a video tube is complicated. It differs radically from attaching a camera to a telescope and taking a picture of some object in the night sky. To begin with, there is no camera. Instead, there is a series of computer programs that govern the movement of the electron beam, the environment within which the scanning occurs, the collection of the data generated by the scan, and the interpretation of the data to finally produce an image. So, obviously, we are faced with a number of questions. First and foremost is, "how do we know that the image on the screen is an accurate representation of what is actually there?" This problem is compounded by the fact that there are no colors at the nano-level. So, in order for us to "see" what is there, the "image" must be artificially colored. Those bright-red pictures of nano-tubes that grace the covers of *Science* are the creations of some clever computer person writing code for coloring representations of nano-objects. Further, there is no way to determine that the objects depicted actually have the shape and relative size they are portrayed as having. The best we can do is to see if the objects behave in the way they should if they have the properties depicted—but our predictions could turn out well and the objects being manipulated still not be as portrayed; it could all be accidental. Our predictions of the behavior of the planets on an Aristotelian model of the universe were fairly accurate, but the planets turned out to have very different properties than that model suggested.

The problem of determining whether or not the objects presented to us by microscopes and the associated computer technologies is permanent. The best we can do is form some hypotheses, test them and see if our predictions turn out, and then refine our models and try again. We may become so good at this that we may convince ourselves that what we see on the screen is a completely accurate representation of what is actually there—but that would be foolish. We should never allow complacency to substitute for certainty. The fact of the matter is that at our current stage of technological development we

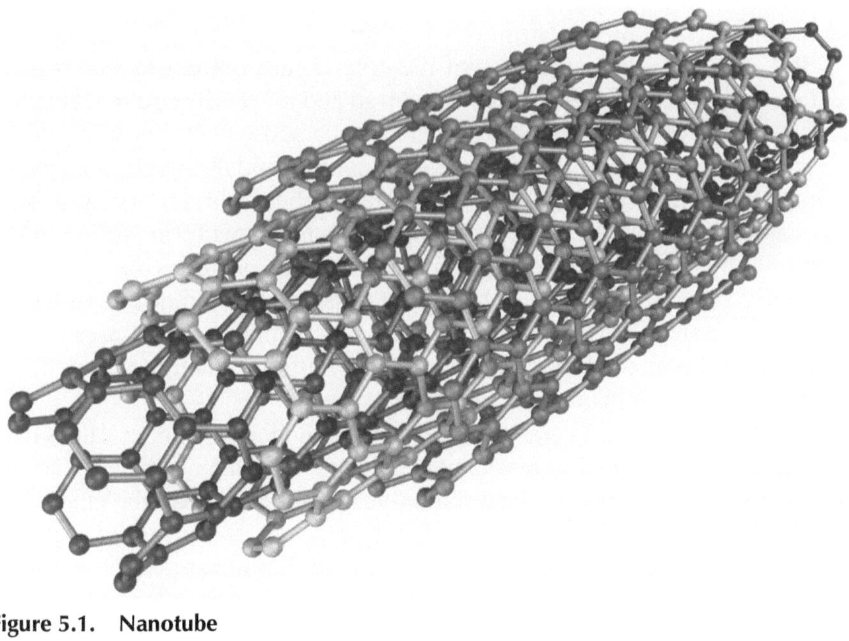

Figure 5.1. Nanotube

cannot know what the nano-world looks like, even if we had some consensus on what it is to know something.

To date, there are few agreed-upon accounts of what constitutes knowledge. The account that requires certainty has to be rejected since we can never be certain about the truth of our empirical claims—things could always be otherwise, so fussing about "certainty" seems to be a waste of time. A popular, but flawed, view claims that knowledge is justified true belief. A justified belief requires an adequate account of "justified," for which we currently have none. For a belief to be true whether or not you know that it is true hardly yields knowledge—for knowledge is more than a string of beliefs.

I advocate a pragmatic theory of scientific knowledge.[1] A proposition, or statement, or claim, P, is considered to be true if when I assume P is true and act on that assumption and my action is successful, then I am entitled to continue to assume P is true. P is never considered true absolutely. Its truth is always tentative, contingent upon being a guide to successful action. Successful action is the mark of knowledge. Successful action is not itself knowledge; it is the mark of knowledge—meaning that we can assume the holder of the beliefs that generated the action has good reason to act on P. What those reasons are is fundamentally irrelevant, but they can be of three types, past-certified empirical "truths," values, and goals. These are the factors that ground an individual's actions. But they are not, singly or in combination, knowledge, for knowledge is not created by nor the possession of a person. It is the creation of a body of researchers, a body that is constantly changing membership. That body determines the criteria for accepting P, *ceteris paribus*. But if over time components of the criteria turn out not to lead to successful action, then the criteria will be revised and there will be a reevaluation of the contents of the house of knowledge and a subsequent house-cleaning. The knowledge creation process is always ongoing, and the results are always in flux. What we have then is a body of researchers actively trying to find out what the world is like and submitting their candidates for inclusion in the house of knowledge to the larger group, which group examines the candidates under the criteria they have developed in the same way and decide to let the candidate in or not. This is the process basically displayed in the peer review process. Whether in applying for a grant or submitting something for publication, the final determinant is the group of experts.[2]

Now this account has its drawbacks. For the individual, his or her values may override the evidence against his or her hypothesis. A creationist may not accept the evidence that this fossil is one billion years old, older than the amount of time the Bible says it took God to create the universe. Likewise for the group, what passes for the values for the group may be the infrastructure they have sunk so much money into that to reject its findings would be to admit to having made a series of flawed decisions based on a set of flawed assumptions.

So, if we bring this back to knowledge of the nano-world, we will find an additional wrinkle. Our access to the nano-world is by way of microscopes. They too will constantly be improved and fussed with, meaning that a reevaluation of older claims following from the use of older STM technology must proceed whenever a new or improved STM arrives on the scene. From my point of view, as a good Heraclitian, there is nothing wrong with the state of our knowledge being in constant flux, for the alternative is a house full of certainties, never to be challenged, with no more room in it once full, which puts an end to research. That we proceed in our efforts to uncover nature's secrets with potentially false background knowledge, using the potentially misleading technologies we have at the time, hoping it will work out in the long run, seems to me just plain common sense, hence my common sense pragmatism. It is sometimes claimed that science is our best source of knowledge, but the history of science shows us that all previous theories have been false, leading us to the conclusion that all current theories are most probably false as well.[3] But that does not mean that we should stop trying to find out what is out there and how it functions.

There are lots of reasons to be happy with the constantly changing contents of the house of uncertainty. To begin with, accepting that our current scientific knowledge is most probably incomplete, if not false, spurs further research, with its concomitant demands for improved technologies. Second, accepting that our knowledge is incomplete or maybe even false should generate a sense of humility, and a sense of humility can block the hubris of assuming we know how it all works. Third, we live with uncertainty on a daily basis, even if we choose to ignore it and press on as if we knew what we were doing. Accepting uncertainty can lead to less neurotic assaults on nature. And this is also what fuels the search for new and better instruments, techniques, theories, and infrastructure. In the long run, the question "Why does this work the way it does?" undergirds all research, and it proceeds not from certainty about the results, but from an awareness of our limitations and our attempts to overcome them. Ursula K LeGuin seems to have arrived at the same conclusion in her science-fiction novel *Left Hand of Darkness*:

> "The unknown," said faxe's soft voice in the forest, "the unforetold, the unproven, that is what life is based on. Ignorance is the ground of thought. Unproof is the ground of action. If it were proven that there is no god, there would be no religion. No handdara, no yomesh, no hearthgods, nothing. But also if it were proven that there is a god, there would be no religion. . . . Tell me, genry, what is known? What is sure, unpredictable, inevitable—the one certain thing you know concerning your future, and mine?" that we shall die.
>
> . . . yes, there's really only one question that can be answered, Genry, and we already know the answer. . . . The only thing that makes life possible is permanent, intolerable uncertainty: not knowing what comes next.[4]

There is an additional wrinkle when we talk about knowledge of the nano-world: what we are presented with on the computer screen is an image, but what kind of an image is it? To accept the claim that using microscopes we can see what is going on at the nano-level will force us to expand or change our understanding of what it is to see something and, in this case in particular, to understand what constitutes an image.

It is a commonplace that the meaning of words changes over time. They do this in a variety of ways. The class of objects the word or phrase refers to may expand. Thus, the meaning of "men" in "All men are created equal" has expanded to include women, African Americans, and other minorities. However, we often do not pay attention to the fact that, while we continue to use a word whose meaning we think we understand, in this case "see" and "image," we also sometimes extend the meaning of that word by applying it to novel situations that only legitimately apply, at best, metaphorically, as I argue later. Eventually what is at first a metaphorical extension of a term may become an accepted part of its meaning. Nevertheless, we should be sensitive to the fact that the meanings of words change over time. This claim is part of a more general thesis about what we do when we employ novel instrumentation. We often employ words whose meanings we already understand in an effort to characterize the sort of thing we think we are doing with this new instrument.[5] This is despite the fact that seeing through a microscope is not the same as opening one's eyes and seeing a tree in front of me, if we were to adhere to a strict sense of "seeing." In extending the meaning of words metaphorically, we also change the meanings of the members of the family of concepts with which they are associated, such as explanation and evidence. Extending the meanings of words results in extending the set of inferences you can make, hence new claims regarding evidence and what counts as an explanation. The metaphor reconfigures the concepts, which changes what we know.

These sorts of metaphorical extensions happen often. In a recent issue of *Time* on the potential of the James Webb telescope, due to be launched in 2020, we find the following,

> Altogether, the assembly has seven times more light-collecting space than the main mirror of the celebrated but aging Hubble Space Telescope. So big an eye will give the Webb the power to look much further into space—and much further back in time—than Hubble can.
>
> That might reveal something spectacular—possibly the very moment in cosmic history when the first stars switched on. "We will be watching the universe light up after the Big Bang," says NASA's Eric Smith, Webb's program director. (Time Magazine, July 3, 2017, p. 64)

The problem here is that the Webb is not an optical telescope. It is an infrared telescope. The "light" it will collect we cannot see. But here we have Eric

Smith talking about "watching the universe light up"—hyperbole to be sure, but we also won't be watching anything—our computers will be translating.

If we take Rasmussen seriously, what the electron microscope does is produce an image. But this is unintuitive for the reasons given later. Furthermore, to claim that an image is produced, suggesting by that that the image is a genuine and realistic representation of what is really there, has serious ethical and social consequences. I will discuss images first and then turn to some of the disturbing consequences of thinking about "seeing" by way of an STM.

Imagine, if you will, a very accurate tennis ball machine. It is a device that shoots tennis balls at you so you can practice returning them without having a human serving partner. Let's assume you aim this machine at a wall built from rough-hewed stone. Your job is to construct an accurate representation of the surface of the wall simply by observing the directions of the balls as they bounce off the wall. Clearly, you will need some help to do this. You need to know a lot about the physics of objects colliding and how irregular surfaces change vectors and so on. You will also need to know a lot about translating what you see happening to the balls shooting off in this direction and then that, and the effect of the texture of the surface of the wall. It is not as if you are directly drawing what you see when you look at the wall. You are interpreting the action of balls as indicating something about the surface and then you are putting your guesses down on paper. That, with minor modifications, is what the alleged image produced by an STM is supposed to have accomplished. But instead of a person doing the drawing, a computer and a computer program, or several, do it. And we are asked to consider the result an image of the surface. Take your hand, if you will, and run it over your shirt. Now draw what you felt. It is not easy, is it? That leads me to ask, *when is an image not an image?*

To answer this question, we should begin by trying to figure out what an image is. This is not as easy as one would suppose, for we tend to use a substantial vocabulary of what we often incorrectly take to be more or less synonymous terms when talking about what microscopes produce. Thus, there has been a lot of loose talk about images, representations, and others. Terms like these are often used interchangeably, mangled, and generally semantically violated. I do not claim to offer much of an improvement—but, at least, I want to alert us to the problem of "image-talk." In cases like this, my preferred method is to work our way toward a common sense understanding of what ought to count, in this case, as an image.

I begin with the intuition that an image is a representation—where a representation is the result of an attempt to capture the salient features of an object, scene, state of affairs, idea, and so on. Fortunately, or unfortunately, what constitutes a salient feature is a function of the person or persons, or technology constructing the image. As a first pass, consider the following items as images:

- Sculpture,
- Photographs,
- Portraits,
- Still lives,
- Landscapes,
- Various kinds of drawings, including computer-assisted renderings,
- Motion pictures—both animated and realistic,
- Visualization inspired by poetry,
- Visualizations inspired by music,
- Plays,
- Operas, and
- Ballet and interpretive dance.

If we accept the fact that these are images, then a Picasso such as the *Guernica* counts as an image, but it would also seem that a Jackson Pollack would not, in so far as it is unclear what a Pollack is supposed to represent.[6] This entails declaring that to be an image is to be representational. But it says nothing about what makes something representational. That said, nevertheless, it is not shocking to note on this account that not all paintings are images, where a painting is nothing more conceptually complicated than paint deliberately applied to a surface. But if it is true that not all paintings are images, especially when they are not representational, have we not found a way into our question, when is an image not an image? It looks like we could reasonably say that an image is not an image when it is not representational. On the other hand, doesn't that just beg the question? After all, it isn't clear that for an image to be an image it must be an image of something. When you think about it, on the one hand, it seems arbitrary to demand that images be representational, but on the other hand, to do so seems to beg the question. For example, consider the following as candidates to be added to the preceding list:

- Diagrams,
- Flow charts,
- Data charts, and
- Spreadsheets.

The interesting feature of these sorts of things is while they are not representational, they do convey information in visual form, for, on the surface at least, it seems as if these forms of images have a different semantics than written language. The important point, however, is that they do seem to have a semantics, for they do manage to convey information. So, if we accept this approach, then one answer to our question is that an image is not an image when it is not representational but conveys information nonetheless.[7] With the reader's permission, let us accept that for the time being as a first pass.

However, that just moves us back one step, for now we can reask the question that our quick look at electron microscopes motivated: when is an alleged representation a representation? The point here is epistemological.

It does not seem too radical to suggest that seeing is a complex activity in which we have to learn to see that there is a car there and we then forget we had to learn that. In our mature state, we see the world around us and we see it for what it is, more or less. That is why philosophical questions like "but are you seeing what is really there?" seem so silly. But, on reflection, we also understand that seeing is partly an interpretative process and that we bring to our seeing a load of background information and experience. I have already argued that to call a seeing of images generated by an electron microscope "a seeing" is to employ a metaphorical extension of our common sense notion of seeing.

There is a lot involved in appealing to metaphor here. If we unpack it, as I would like to start doing, we can see that to understand through metaphor is to do a number of things at once. First, we use metaphor to access what is new and different because in a metaphor we take what we know and apply it to the unknown and say that the unknown is like the known in these various ways and unlike it in other ways. The use of metaphor makes the new seem familiar and approachable, usually. Sometimes, as in the example of the tennis ball gun discussed earlier, it makes the unknown or the new seem even more strange than we first thought. Second, when using metaphor to make the new and unknown approachable, we are also asked to accept that certain things that we really don't know are, nevertheless, reliable. Metaphors tell you this is like that in certain limited ways, and, by the way, just accept that everything else is working just fine, however that happens. In the case of the electron microscope, when asked to accept what it produces as a representative image we are also being asked to accept the fact that the assumptions built into the manner in which that image is constructed are correct and reliable. To use the language of science studies, we black-box the process and merely look at the result. But to call the image created by the electron microscope an image is to ask us to accept in some fundamental way that the science is sound and the technology along with the programming is reliable and the people manipulating it are honest.

But it seems this is a lot to ask. What is interesting is that it appears that it is not a lot to ask. It is a measure of the success of the scientific establishment that we, the general public, tend to accept claims based on the use of increasingly complicated instruments working in the realm of the frontiers of scientific research with increasing readiness. That is, the more complicated the science and the more simplified the public explanations, the more readily we tend to accept those fantasies. That is why it is important to know what really happens in an electron microscope before buying into the claims with which it is associated. Before I explore what that ominous remark is supposed to suggest, consider one example of the phenomenon to which I am referring.

We are all in awe of the images sent to us by the Hubble Space Telescope. The ones of the Horsehead Nebula and the crab nebula are just breathtaking—and the colors are truly inspiring—just one catch—the colors are computer generated. When I tell my students that their faces resemble the ones they have when they learn, there is no Santa Claus.

Computer-generated, color-enhanced images of the nano-world are pervasive. One of the best known of such images is the picture of the letters IBM.

Figure 5.2. Horsehead Nebula

Figure 5.3. Crab Nebula

It was presented as evidence of our ability to manipulate pieces of matter as small as atoms to produce such an effect. The problem is that there is no color at the nano-level. Even the decision to present the IBM picture in gray tones was a decision to create an image in a certain way. So when we say of an image that it must convey information, should we not also be asking (1) whether this is a claim that reality is being represented and (2) is the image presented of something real or imagined? Perhaps, then, should we not be asking this slightly different question: "When is an image not an imagining?"

The issue here is both epistemological and ethical. The epistemological issue concerns, for lack of a better term, noise. We are familiar with the problem of filtering out noise when searching for an identifiable signal. The problem is multifaceted: what to filter out and, on what criteria, to amplify, to what degree, and so on. The problem with color-enhancement and sharpening up of nano-images is that we don't know which elements are important and which are not. Further, the problem may be intractable since we do not have a god's eye view from which to determine if we have it right. In a certain sense, the problem here is in-principle-lack-of-access, or to put it differently, a case of very strong underdetermination. But is this really a problem? We have in-principle-lack-of-access to many astronomical events, like the big bang, and we still claim to know a lot about the early moments of the universe (and, if the Webb telescope does what it is anticipated to do, we will soon know even more, or will we?). We have images from the Hubble of far-distant galaxies that we can never get close to in person, and yet we can still understand a lot of what is going on there—or so we think.

My worry here is that, unlike the "images" from the Hubble, we have relatively little experience in enhancing images produced by microscopes. We have ways of checking up on the Hubble images. For example, we can experiment with filters and use smaller telescopes here on earth to check out their effects when we look at mountains or trees. However, although we have lots of experience with so-called images from STMs, we do not have such successes in fixing them up. This is, in a curious way, a new version of the what-are-we-going-to-do-when-we-stain-a-specimen-that-we-are-going-to-examine-under-a-standard-microscope problem. Computer enhancement of images can be fun, especially with all the nifty colors we can use. But is it producing an honest replication of the object/surface in question? Clearly, we cannot answer that question, and that raises the ethical issues. But before we go there, we need to consider an objection.

The statement "Clearly, we cannot" might be considered contentious, but with a bit of explanation it should be obvious. The surface on which nano-scale objects exist is at the interface between the quantum domain and the atomic. We have no idea how to visually represent what happens in the quantum domain. In fact, because of the behavior of objects in that domain, we cannot know in principle what is really going on, so we cannot say we are accurately representing the surface on which the atomic structures we are picturing sit. If we cannot claim to be accurately depicting the surface, then how can we be sure of the space in which the nano-structures function? If that is uncertain, so must be our representation of the nano-structures themselves.

Turning to the ethical issues, they come in two forms: strong and relatively minor.

The relatively minor issues have to do with the relationships between the scientific community and the general public. For example, we are misleading the public when we fail to disclose fully what we are doing when we enhance our electron microscope–constructed images using a computer. And it is not enough to stick a "computer-enhanced" phrase as a disclaimer under the image.

The strong ethical issues center around the fact that these images raise false expectations. Among them, that we know more than we do. The presentation of these beautiful pictures suggests, in a very strong way, that this is indeed what it is like out there or in there. But, more importantly, they mislead in other crucial ways. The beautiful computer simulations we see of nano-interactions are not only beautiful simulations, but they are also almost heart-stopping in their ability to feed the hubris we sometimes exhibit when employing the newest technological toys, computer and advanced programming techniques among them. Do not get the wrong impression—I am not suggesting that we should not employ the latest technologies when doing scientific work. What I am talking about is the illusion we create not just for the general public but sometimes in the practicing scientific community. The illusion is that we know more than we really do. Never underestimate the ability of human beings for self-delusion.

These computer-generated and -enhanced pictures suggest that the world is, at rock bottom, a simple place. It can be pictured as individual atoms resting on stable fields that we can manipulate at will, twist them, enlarge and narrow them, put them to music, make them dance, when in fact nothing of the kind is the case. On the contrary, the world at the nano and quantum mechanical level is buzzing, shifting, constantly in motion in a nonlinear and nonclassical causal fashion mess.

This is all heading in one direction. It is not just misleading to suggest that the world is simple at the "bottom." It is epistemologically suspect. It employs a crucial, but faulty, assumption. It is the assumption that the world is better understood if we simplify our presentations of it. This is fundamentally wrong-headed. It may, in fact, be helpful to extract some feature of the world, color it with pretty non-natural colors, and play with it. But it is more important to put that heuristically altered item back into the buzz and to try to understand it in that environment, its "natural" environment. Most importantly, it is crucial that we explain to the public and to our colleagues the purpose of the heuristic move and what it reveals about what is really going on at the bottom.[8]

So, what is wrong with simplification? It suggests that we know more than we do and, crucially, that we can do more by way of manipulating nature than we can. The scientific community has done a good job of convincing the public that it has god-like powers—but this situation presents a double-edged sword; the public feeds on gods that fail. Be honest about the mess, and you will reap positive rewards. Further, it is not the simplicity of the universe that makes it the object of our enquiry; it is the complications, the unanswered questions, the mess of it all. The more we look, the more complicated we find it to be, as in the unexpected discovery of a new particle at CERN in the fall 2018, thought to be twice as heavy as the carbon atom. If you coddle the public and give them simplicity and then in the crunch, when, for instance, in the hospital, you say "well it is more complicated than that," then you will have failed miserably.

I love the pictures, but they are not representations. They are heuristic imaginings, extended metaphors, if you will, and they should be recognized as such and treated that way. How will that affect the way in which the work of scientists is perceived? My guess is that it will enhance it. Doing science is hard work. The public should know that, and when they do the successes of science will be all the more appreciated. Telling the truth is hard.

The question we have been examining is, "in what sense is an STM-computed-generated picture of nano-structures an accurate representation of what is there?" Following some discussion of how "seeing" using an STM involves a metaphorical extension of the concept of seeing, it was argued that to be a representation an image must convey information. The problem, we discovered, is in understanding what the "information" is that is being conveyed, since we cannot directly access the domain that we are purporting to represent. The problem is not that we do not know how to interpret what is presented to us as an image, but, rather, that we have loaded the creation of the representation ahead of time without being able to know if our guess that this is what the STM and its fellow traveler computer programs are producing is an accurate picture of what is really there. The reason why there is so much discussion of when an image is an image is that this really is a question of whether or not the image that is produced is an accurate portrayal of something that is really there or a mere fabrication, a creation of the technology itself.

We explored the use of metaphors as a possible end run around the epistemological problem of limited access to the domain of enquiry and discovered that it too has its limitations. The primary limitation is that for metaphors to work we must assume that there is a high degree of symmetry between the known domain we are using as the basis for our metaphor and the unknown domain. But we cannot make that assumption, for to do so is to beg the question.

Consider one last attempt to convey a sense of the magnitude of the problem. If we do a random sample of some domain and then plot the results in three dimensions, assuming that that sample is truly random and there is no natural clumping of the data, which curve is the correct one? We can draw an infinite number of curves through the data. Without an independently certified decision procedure for selecting the correct curve, we are simply left with the data. The problem is further complicated by the fact that there are ethical dimensions. (1) To say that this is what is taking place at the nano-level is to lie, since we don't in fact know that to be the case. (2) To present these standard, nicely colored, enhanced, and simplified pictures as genuine representations of what is going on at the nano-level is to claim falsely that nature is in fact simple and clean and neatly colored at the nano-level as it is everywhere else. But nature is not neat and tidy at that level or anywhere else. We *impose* order on what we see. To suggest otherwise is to mislead by way of making it appear there are simple answers to very complex problems. This approach gets us into trouble at the political level, and it should get us into equally a big trouble in our epistemology.

This move to simplification manifests itself in other ways. Take, for instance, the issue of deskilling of computer imagery.[9] Those pretty images are created by software systems that read the data for microscopes and AFMs (atomic force microscopy). When the STM was first created, the data were in the form of numbers that then had to be graphed and mapped onto topographical scenes. That required human intervention. Today, the data are read by computer programs, eliminating the human, and eliminating the skill of reading the data to make it meaningful. We now go straight from data to picture *via* computer. It is doubtful that contemporary researchers can even read the numerical data today. The universe is not a simple place. That is what makes it so fascinating. The more complicated we discover it to be, the more we will have to rely on our technologies and their infrastructures.

NOTES

1. I am addressing scientific knowledge only here—what I call everyday knowledge relies more heavily on day-to-day experience and is easier to check.
2. What constitutes an expert is a topic for another day.
3. I realize I am committing an inductive inference here and I should be reprimanded. But the histories of the sciences are an awfully large sample class to work from.
4. I owe Ashley Shew thanks for finding this lovely quote. And lots of other help with this project.
5. For an excellent extended study of this phenomenon see I. Bernard Cohen's history of the use of the word "revolution" in Cohen (Revolution in Science, 1987).
6. If turning to art is seen as somehow cheating, it is important to remember that the creation of images began in art.
7. Yes, "information" is not defined. But, I suggest, we have to start somewhere. If we succeed in making progress by proceeding in the manner suggested, we can always return and fine-tune the argument by going deeper into concepts like information. Call this approach "conceptual boot-strapping."
8. The use of the phrase "at the bottom" and its associated phrases in itself presents a curious view of the structure of the universe—as if there is a bottom-most level of reality out of which everything is created, level by level—very positivistic and unjustified. More on this in chapter 7.
9. I thank Ashley Shew for bringing this to my attention.

Chapter 6

Some Consequences of a Technological Infrastructure

The Heraclitian philosophy of science advanced here takes its starting point, as noted in chapter 1, from the observation that the processes and products of science are constantly in flux. We have looked at some of the causes for that restlessness, focusing on the development and role of novel technologies. In chapter 4, we discussed the technological infrastructure (TI) of science, emphasizing what is needed to conduct science in the contemporary world. There are consequences that stem from the embedding of scientific work in this infrastructure, and that is what this chapter is concerned with. The primary issue is the paradoxical result that while science is undergoing constant change and scientists need its technological infrastructure to pursue their goals, it is also the case that the more complicated the TI becomes, the more it becomes an impediment to change. When results showing that the current theory, which relies heavily on its TI, must be either ignored, explained away, or accepted, we need to decide what actually needs to be acted on, parts of the TI or parts of the theory, or the whole theory. If the anomalous (in Kuhn's sense, see *The Structure of Scientific Revolutions*, 1962) results are accepted, the question now becomes how much of the TI needs to be abandoned when we abandon the theory that it supports. That is, we need to be able to determine how much of the theory's success is to be credited to its TI in order to determine how much of the TI we retain as we turn to other objectives.

Another way to put this is to ask, how much of what we thought were positive results of this particular TI/theory amalgam is a product of the TI itself? And to treat this question honestly we have to narrow the scope of the TI here. I suggested in chapter 4 that social institutions like the United States National Science Foundation (NSF) are both an enabling technology (by providing funding) and a directional technology pointing us down specific roads of investigation. The NSF not only funds research, but it also puts out calls for proposals in specific areas, thereby setting a research agenda. It also just may be the case that it takes us down the wrong roads, that is, someone's great idea that turns out to be empty—for example, cold fusion. So, money and time are lost, but maybe no direct harm is done. Where there might be damage is when a new theory at first produces good results, allowing us to invest more money in the TI, only to discover down the road that the theory is flawed in serious ways. At that point, we are left with an expensive TI and no theory.

There are many ways a theory can be flawed. Some of its initial assumptions may be wrong—that is, there is no such thing as phlogiston. While at first blush the structure of the theory may seem productive, the consequences of applying its results in conjunction with those initial assumptions and the structure of the theory may yield false results. Despite the best efforts of the logical positivists, there is no one preferred structure that a good theory has to have. So, it is impossible to know ahead of time if the theory is structurally sound. But there are ways to find out where we went wrong.

Let us begin by asking how much of the confirmed content of the theory is a creation of the instrumentation that was used to reveal it. This was a charge leveled against Galileo's claim that he discovered satellites rotating around Jupiter. The images, his opponents charged, were merely the creation of the flawed instrument. And to some extent the charge was not improbable since the quality of the glass used in the telescope was quite poor.

The second question to pursue is to ask, Where is the theory weak? There are several possible weaknesses. To begin with, the theory's fundamental assumptions could be flawed. These might be found in the nature of the unobservable or theoretical entities it proposes. Thus, Aristotle's assumption that earth, air, fire, and water were the fundamental elements was a mistaken assumption. The theory may also lack explanatory success. A major flaw in the Aristotelian/Ptolemaic account of the movements of the heavens was its inability to account for the apparent retrograde motions of the planets. Fixing this weakness required a number of *ad hoc* assumptions, thereby weakening the theory's explanatory power. A third possible weakness of a theory can be that it has the relations among its fundamental assumptions wrong. Let us assume for the moment that Einstein had it wrong when he asserted that $e = mc^2$. What if the correct equation were $c = me$?[1]

If, in fact, it is true that the history of science is the history of failed theories, we are caught in an epistemological bind. There is a certain sense in which we know deep down inside that the current theory we are defending is false, but yet we must proceed as if it were true because to do otherwise would be to allow us to be sloppy when devising tests for the theory. As in "let's do X, even though we know it is intuitively implausible, but that doesn't matter since the theory is in all probability false." But this is exactly what KATRIN is all about (see chapter 7).

Returning to our TI, we are in a similar epistemological bind. Here is how the picture looks: having committed ourselves and having convinced our funding sources, we should be encouraged to work with the latest, most expensive, and technologically complex technologies; we literally can't afford to throw it all away. To do so sends a message to our funding sources, something to the effect that we don't know what we are doing. If that is the perception, why should they fund us again when we submit a new proposal, asking for expensive new technological goodies to fill up our sandboxes?

The solution to these various binds is to follow Feyerabend (see *Against Method, 1970–1975*, p. 156). "Anything goes" is a methodological injunction to do whatever it takes to defend your theory, and I will extend that to defend your TI as well, because to do otherwise is to not seek the truth in good faith. In short, to be a good Heraclitian is to commit to using every trick in the book to defend your views, while at the same time knowing that, in the long run, they will be shown to be false. But knowing that in the long run they will be shown to be false shouldn't be a problem since you have to work with what is available at the moment, knowing full well that down the line new ideas and new technologies will be coming along that will allow the future researchers to remark on how primitive your science was. If you don't commit to doing everything you can with the means available to you, you are looking at the "can't get started" argument. That is, if knowing your work will be superseded, then why bother in the first place? Well, your work (and by "your" I mean the work of everyone in your research area using the same TIs and theories) is needed as a stepping stone to the future. While the history of science is the history of failed theories, investigating those theories was needed to get us where we are today. Look at it as investing in the future.

The good Heraclitian understands what it means to be committed to the view that everything is in flux. First and foremost, it doesn't mean that everything is changing at the same rate at the same time. Some changes are slower to develop, but that doesn't mean we should be fooled by the appearance of permanence. Second, showing that some TIs and theories don't give us the results we hoped for is not to be understood as having a license to reject the entire enterprise. It does provide us with the right to ask questions about what worked or which insights we derived we can keep. For example, rejecting the view that the earth was the center of the universe did not cause us to reject the idea of a center, not at that point at least; we required the development of a new TI relying on massive space-based telescopes and other instruments to give us a better "view" of where we stand in relation to everything else. Finally, operating with the knowledge that what we discover about the universe most likely will be shown to be false at some point in the future is no reason to stop doing what you are doing. We should know by now that we operate with incomplete knowledge and inadequate TIs, but the aim is not to give up, but to strive to improve both. Yea team! Fight, fight, fight!

Returning to the serious, the short take on what a Heraclitian technological philosophy of science looks like is very much like common sense. Individually we make huge investments in material goods in our daily lives, automobiles, houses, gadgets, and so on. We know better than to assume that since we paid a lot of money for that car, it is going to run flawlessly, but we use it assuming that it will, knowing that it won't at some point. We don't let the fact that it will break down sometime, somewhere, stop us from getting in it and driving to Alaska. We take precautions such as having it regularly serviced and buying new tires when the old ones get thin. We wash and clean it to keep deterioration to a minimum, and when the time comes, we trade it in on another expensive vehicle and start the whole cycle all over again.

If we were to live our lives acting only on that about which we could be absolutely certain, we wouldn't live our lives. We rarely, if ever, stop to think about the complexity of the infrastructure we operate in on a daily basis. Sometimes, it is brought home to us in dramatic fashion. For example, I was raised basically as a city boy. But after graduate school, when my spouse and I moved to Virginia Tech, we shortly moved out of the town of Blacksburg to a farm in the country. The first time the power failed I freaked out.[2] Because we relied on well water, we had no water. We had no lights. We had no news. We were totally isolated. The first thing I did was to buy a battery-powered radio! And then we took matters into our own hands and began to construct our own infrastructure. We purchased a generator, so we had water and could run a few appliances. We bought a tractor, so we could reclaim the pastures. We put solar panels on the barn roof. In short, we created a technological infrastructure of our own that allows us to live in isolation when the more developed technological infrastructure fails.

The basic point here is that we live and manage uncertainty on a daily basis—and work to anticipate what we will need in its face. To cope with that uncertainty we build a technological infrastructure to stabilize our environment to the extent possible. Why should we do otherwise when it comes to dealing with the technologically immersed science we engage in? Furthermore, we engage in two important activities in both daily life and as practicing members of the technological infrastructure of science, assessment and feedback loops.

When something goes wrong, we generally tend to ask why it went wrong. And when we find out, we try to determine what we could have done to avoid the failure, that is, we assess the situation. Having assessed the situation we try to determine what assumptions we were operating on that needed adjusting. These assumptions fall roughly into three categories: knowledge, values, and goals. Did we assume we knew something we didn't? In the case of our car example, did we ignore the warning signs that the battery was failing? In the case of our values, did we try to avoid spending the money to replace the battery because we hate to spend money on the car? In the case of goals, did we really want to be able to brag about how long our battery lasted? Having determined what assumptions we were operating on that got us into trouble, we take that knowledge and go back and fix the problem. That is using a feedback loop. This framework is helpful because it allows us to schematize the common conception that if there is a scientific method it is that science proceeds by trial and error. It is also helpful in showing how common sense and scientific reasoning are on a continuum. We reason the same way in our daily lives as we do when solving an engineering design problem or a scientific problem. This approach also opens the door to addressing the basic component in the conduct of scientific activity within a technological infrastructure, the people doing the work.

When things go wrong, we tend to blame it on something like a flawed piece of equipment or a faulty assumption in a theory. But we rarely identify personal decisions as responsible for a technological/scientific failure. Two examples help point this out, the *Challenger* disaster and the initial failure of the Hubble. In the case of the *Challenger*, we know the engineers on the project warned that the rings would not function properly in the cold and that their warnings were ignored for political reasons. The case of the Hubble is more complicated.

How did NASA manage to launch a defective telescope? Or, not to appear to put the blame on NASA, how did a defective telescope come to be launched?[3] The U.S. Congress, following the discovery of the blurry images coming back from the Hubble, ordered an investigation. This resulted in a report presented to Congress by William Colvin, inspector general of NASA. Several things are not in dispute: (1) The primary mirror of the Hubble Space Telescope was flawed. It suffered from a spherical aberration. (2) The flaw was the result of a manufacturing error and several failures on the part of management. (3) The manufacturer of the mirror, following the investigation, agreed to repay the government $25 million. The willingness of the manufacturer, Perkin-Elmer, to agree to this settlement, suggests something more than a mistake was involved.

The problem was that the main mirror of the Hubble suffered from a spherical aberration. According to Eric Chaisson:

> Spherical aberration is an undesired optical condition induced by a circular mirror having a deformed figure—that is, a curvature of the wrong "prescription." It has nothing to do with the smoothness of the mirror, but depends only, yet critically, on the shape of the mirror. Unable to focus all the light to a single point, such a misshapen mirror actually displays a range of foci. Alignment of

> the optics can emphasize the light captured either by the outer or by the inner parts of the mirror, but . . . the resulting image, regardless of these attempts, is a bright point surrounded by a disk of diffuse light. (Chaisson 1994, p. 174)

The revelation that the telescope suffered from this specific problem and not from a failure to focus the primary and secondary mirrors was first proposed by scientists at the Space Science Institute. The NASA engineers and the engineers at the company that manufactured the mirror, Perkin-Elmer, initially objected and sought to find the solution in adjusting the two mirrors for focus. After all efforts were exhausted, the engineers agreed to the diagnosis of a spherical aberration.

This was an expensive project, paid for by taxpayers' dollars—so naturally there had to be a congressional inquiry. The result of the House inquiry was an extensive investigation with a report from NASA following. The investigation revealed the following six significant irregularities (and here I will quote directly from the testimony of Bill Colvin):

1. **Non-approved Reflective Null Corrector Washers.** In the process of adjusting the spacing of the reflective null corrector for the HST Mirror, technicians discovered they could not move the field lens into the prescribed position. Instead of calling in the designer, the *contractor* inserted an ordinary washer under each of the bolts holding the field lens retainer to the adjustable plate. The insertion of these non-approved washers in an instrument whose precision is measured in tens of millionths of an inch required a nonconformance report. There is no written evidence that a report was generated.

A nonconformance report is a report describing how what was done deviated from the specifications of the government design. Why did the Perkin-Elmer engineers fail to file such a report? In an article in *Science*, the explanation rendered is that officials at both NASA and Perkin-Elmer

> allowed themselves to be overwhelmed by the massive cost overruns and schedule slippages in other parts of the project. As a result, they neglected the mirror work, which seemed to be relatively well, and failed to enforce their own quality assurance procedures.

We can understand how this situation developed by concentrating on why the decision was made not to call the designer back in: cost overruns and schedule delays, externalities with a major impact on the final product. My emphasis is on the decision-making and the factors that bear on it. In the remaining five items in Colvin's report, each point comes back to the decisions made by the individuals involved. The factors playing into those decisions varied, some of them were internal to the firm, some were not. However, only by directing attention to which people made what decisions with respect to the materials used and the objective to be obtained can we understand what happened. Unless it requires commentary, I will just go through the last five findings of the investigation.

The second item Colvin relates is as follows.

2. **Unexpected Results from Inverse Null Corrector.** The inverse null corrector was used by Perkin-Elmer as a part of the reflective null assembly. It could be used as a check on the reflective null corrector's alignment and stability. It could show whether there was a gross flaw in the reflective null corrector as well as measure its stability. Since the inverse null emulated what the primary mirror surface should be when finished, it should have produced an interferometric pattern with straight lines, or in other words a null condition.

   Instead, a pattern of wavy fringes was produced by the inverse null corrector at each testing cycle using the reflective null corrector. Although the inverse null corrector was producing results which did not agree with expectations, no nonconformance report was generated, and the designer was not consulted. While no qualitative analysis was performed, the inverse results were dismissed as being attributable to error in the null corrector's manufacture.

   As a result of this condition, *Perkin-Elmer managers* decided to make a design change which revised the usefulness of the null corrector from its initial utility as a "double check" on the health and stability of the reflective null. Through testimony we ascertained that *Perkins-Elmer managers* decided to use the inverse null corrector solely as a stability check.

3. **Refractive Null and Reflective Null Tests Do Not Agree.** Within a week of the first reserve null results, Perkin-Elmer received other unexplained testing results. A refractive null corrector was used by Perkin-Elmer in Wilton, CT, for initial grinding of the mirror blank to get the mirror surface shape roughly to the desired finished shape. The mirror was then transported to Danbury, CT, where testing was first conducted using the reflective null corrector.

   The test results from the first reflective null corrector test and last refractive null corrector test did not agree. Yet the expectation was the interferograms would be similar. In assessing the differences, Perkin-Elmer commented on the poor quality of the interferograms—caused by the rough surface of the mirror. They also asserted that some hand polishing of the outer edges of the mirror had occurred after the mirror left Wilton. No further quantitative analysis was performed.

4. **Recommended Gross Error Test Not Performed.** At the conclusion of the polishing phase, a Perkin-Elmer Vice President and General Manager formally requested an internal review of the primary mirror certification by senior scientists employed by Perkin-Elmer. This group, called the technical advisory board, held a review of the mirror fabrication and test results. On May 21, 1981, they recommended to the Perkin-Elmer Vice President that an independent test be performed on the mirror. "The recommendation stated that another test of the figure using an alternative method such as a Hartman test or a bean should be made". The purpose of the test would be to uncover some gross error such as an incorrect null corrector. No such test was ever performed.

If no such test was made, we can infer that the Perkins-Elmer Vice President decided not to make it. Why he decided that most probably has to do with the factors alluded to above, cost overruns and schedule delays.

5. **Vertical Radius Test Anomaly.** In May 1981, the most significant of the irregular events occurred when the refractive null corrector was used to measure the center of the curvature of the mirror. . . . the refractive null interferogram showed wavy lines—clear evidence of an error in one of the measuring devices and possibly the mirror. . . . the chance that two separate measuring devices, the inverse and refractive null correctors, would find matching errors and both be wrong and the reflective null right—is infinitesimal. . . . Yet Perkin-Elmer. . . . failed to resolve the discrepancies in a quantitative way. . . . . had Perkins-Elmer attempted to determine the source of the error, with the analytical and measurement tools in place at the time, they could have determined that the flaw was in the reflective null corrector in one or two days. *Perkin-Elmer personnel* assumed erroneously that there were large "as built" errors in both the refractive and inverse null correctors. They ignored the results of the intended sanity checks on the reflective null corrector— . . . the results of the last . . . test concerned Perkin-Elmer managers, but they did not disclose the results or their concerns outside of the optics fabrication group. . . . to our best determination, Perkins-Elmer did not share the discrepant results of the vertex test with NASA. The NASA Plant Representative . . . was provided a copy of the center portion of the vertex radius test interferogram. This cropped version did not disclose the curved fringes which would have indicated a problem with the test results.

Here we are at a crucial point. Someone deliberately cut out the center of the interferogram in order to not reveal evidence of a problem. The decision to hide relevant evidence is at the heart of the resulting problem with HST, since, as we shall see, other members of the Perkins-Elmer group were left in the dark. Colvin's report continues:

> Perkin-Elmer quality assurance personnel told us that they were present at daily meetings on the HST and they were never made aware of aberrant test results, nor were there any such discussions at the meetings.

If this last point is true, then it suggests that communication within the company was flawed. It is not clear if there was competition or merely an effort not to look bad. It begins to sound like upper management charged with overseeing manufacture of the mirror dug itself into a hole that further decisions simply deepened.

Finally, there was rapid close out of the Perkins-Elmer optical fabrication team. The program manager had 11 items he wanted to check out, but was not given approval to do so. And in conclusion, "The Head of Manufacturing Optical Analysis stated that he and his manager would have couched the vertex radius anomaly in terms of a 'need to recertify the reflective null corrector.'" This is nothing more than an admission of intended obfuscation, which tends to support my conclusion.

What are we to make of this incredible list of failures to report bad test results, of failure to adhere to design specifications, and failure to adhere to protocols? It might be attributable to the internal ethos at Perkins-Elmer. But it seems there were other factors at work as well. Robert Smith (1989) details the myriad political pressures that also were at work. Economic considerations were paramount. Virtually every decision from the early days through the mid-1980s was made to lower costs. Everyone wanted a space telescope, no one wanted to pay for it. The decision to use the shuttle as the launch vehicle forced further design changes in the Hubble, which aggravated the complexity of the project. At every turn, as we uncover layer after layer, we find that it is the decisions that were made, motivated by varying considerations from political clout to economics to engineering design, that give us the clues to what actually happened. This is sketchy—the case itself is complicated because the technical nature of the design needs to be part of our conversation, as do the political context and the economic situation. For example, budgetary considerations forced a cutback in the number of NASA inspectors assigned to the project. At every turn, we find crucial decisions based on non-uniform considerations that point the way. Finally, we find that failure to assess decisions in light of the larger scope of the project is what ultimately led to failure. Why bad tests were not reported and discussed up and down the line is not clear. Why designers were not consulted when production problems emerged is not clear. What is clear is that the resulting engineering design failure was a failure to utilize and act on a feedback function.

It is important here to pause to consider a possible objection. It might be argued that the launching of the defective Hubble was not an engineering failure so much as one of management. Clearly, the final decisions not to test and to ignore the results of tests that were already on record were a management decision. But the discussion here is not about engineering versus management. It concerns *the design process* that is part of the technological infrastructure. This was a government contract, and it had a number of specific protocols built into it. These included consulting with designers when problems occurred, using the null refractor results as intended and so on. What happened was clearly a breakdown in the design process, a process that included both line engineers and management. To claim that this was all a management problem ignores the fact that it all began with some technicians failing to follow protocol. My concern here is with how the design process actually works as an infrastructure and how it can fail. What we have here is an example of a process that is set out in writing and then ignored by virtually all parties.

Finally, it is also clear that we cannot argue that these were just bad engineers who were responsible for the problems with the HST, that is, unless you build honesty, apple pie and the American way into your definition of a good engineer. These folks knew enough to engage in systematic behavior designed to hide their mistakes and they succeeded, until someone turned on the telescope.

What can we glean from this brief look at this example? If there is a moral, it is this: the main focus of efforts to understand technological projects needs to be the people doing the work. It is their decisions that hold the promise of understanding how what happened happened. True, those decisions will be contextualized by a variety of factors, but few of them can be determined ahead of time except in the most general terms—people, materials, institutions, and so on.—but that tells us little. The devil is in the details; grand schemes seem increasingly less appealing. All we really need to work on is a better understanding of why people do what they do. This requires a clear understanding of the environment they work in, call it the technological infrastructure, which also includes the social, economic, and political realities at play.

NOTES

1. I realize this is nonsense, but use it only for illustrative purposes.
2. Interestingly enough, when you are in a rural community in the South, it is the "power" that goes out. When you are a town or city person, the "Electricity" goes out. Go figure!
3. Most of the material I use here comes from the U.S. Congressional Record (November 16, 1993).

Chapter 7

Scientific Progress?

If, as noted at the end of chapter 5, our technologies and our theories are inadequate to the task of accurately revealing what the nano-world is like, how do we determine when they have improved enough for us to have high confidence that what they tell us is, if not the truth, reasonably close to the truth?

This is not the same question as "What is scientific progress?" Philosophical theories of scientific progress abound. But, as noted in chapter 1, these rarely have anything to do with the way science actually works. (There are exceptions, such as Nicholas Rescher's *Scientific Progress*, to be discussed later.) They are generally theories of rational decision-making concerned with the grounds for choosing between two theories. Mostly the grounds put forth appear to be criteria for determining when one theory is better than another, which looks like it addresses our question of how we decide that one theory is an improvement over another. But these are radically different activities. Showing that this is the case will occupy most of what follows.

I have been interested in the topic of scientific change for 50 years, since I first read Norwood Hanson's *Patterns of Discovery* (1958). But after graduate school, Larry Laudan's *Progress and Its Problems* (1977) came to dominate the discussion as a worthy successor to Kuhn (1962). In that context, scientific change and scientific progress became interwoven or, as I want to put it now, confused. Scientific change is not necessarily scientific progress. In fact, there is no such thing as scientific progress. Science changes, both in its content and its methods, that is not to be denied. But claiming progress is an attribute of science carries a lot of empty baggage with it and puts us on the wrong path to understanding what happens. The notion of progress strikes a much bolder claim than we need in order to appreciate the process of scientific change and overlays values that are unnecessary and presumptive.

To begin with, "progress" suggests movement toward some desired goal. That goal can be regulative or constitutive. I doubt if progress is a constitutive goal of science. But first we should explore the idea that progress is a goal of science at all.[1] As a first crude stab at the idea, it could mean that science strives to obtain a complete account of the objects, processes, and forces that constitute our universe and the various activities that take place in it. Thus, Steven Hawking claims that science aims at a complete description (Hawking 1988). That assumes that such a complete picture is possible. Peirce thought it was. In "How to make our ideas clear" Peirce put it this way:

> The opinion which is fated to be ultimately agreed to by all who investigate, is what we mean by the truth, and the object represented in this opinion is the real. That is the way I would explain reality.
>
> But it may be said that this view is directly opposed to the abstract definition we have given of reality, inasmuch as it makes the characters of the real depend on what is ultimately thought about them. But the answer to this is that, on the one hand, reality is independent of thought in general, but only of what you or I or any finite number of men may think about it, and that, on the other hand, though the object of the final opinion depends on what that opinion is, yet what that opinion is does not depend on what you or I or any man thinks. Our perversity and that of others may indefinitely postpone the settlement of opinion; it might even conceivably cause an arbitrary proposition to be universally accepted as long as the human race should last. Yet even that would not change the nature of the belief, which alone could be the result of investigation carried sufficiently far; and if, after the extinction of our race, another should arise with faculties and disposition for investigation, that true opinion must be the one which they would ultimately come to. "Truth crushed to earth shall rise again," and the opinion which would finally result from investigation is destined to lead, at last, if continued long enough, to a belief in it. (Philosophical Writing of Peirce, in Buchler, J. (2011) p. 39)

This is Peirce's convergent theory of truth—ultimately, if we persevere, we will find the final theory that will give us the total big picture. Furthermore, there is a total, big picture to be had. But this is not an argument. It is a statement of expectations. We should separate two things: (1) Peirce's theory of reality, from (2) Peirce's theory of scientific progress.

For Peirce, the real or reality is independent of what anyone thinks. That is different from assuming that what we say is real is, in fact, real. We can investigate all we want and come up with any number of theories, but that does not change what actually is out there, even while what is actually out there is changing. This is a pretty straightforward idea and has its attractive features. For one, it harmonizes with the view that just because we say it is so doesn't make it so. In other words, it makes reality objective.[2]

For Peirce, the truth is what corresponds to the real. Finding the truth may take a long time, but ultimately, we will find it and we will know it is the truth because all investigators at this final point agree on it. This view is more problematic. There is no guarantee that we will find the truth. Saying so doesn't make it so. Further, the fact that we all agree is hardly a guarantee that we have the right theory. Even Peirce agrees with that.

To make the point as clear as possible, consider that there is no argument that proves we will arrive at a final theory that captures the whole picture. At best, Peirce's account provides us with a regulative ideal; that is, we should be striving to achieve that final theory that captures reality. OK, but we are still left with the problem of proving that it does, in fact, match reality—general agreement is not enough, to assume that it is and to think so suggests we are under a collective delusion or suffering from a colossal case of hubris.

If we look at the other possibility, that progress is constitutive of science, we fare no better. To say that progress is constitutive of science is to say that any scientific work that does not take us toward some final goal is not merely bad science, but not science at all and that is just silly. It flies in the face of a deep-seated understanding of scientific method as one of trial and error. In short, in order to make any sort of progress in science, you have to make mistakes, which doesn't mean that when you do you are not doing science.

There is another monster lurking in the background. It is this talk about "science." Talk of scientific progress brings with it the skeleton of a positivistic conception of a unified science. But, as we have noted earlier, there is no such thing as science *simpliciter*. There are many individual sciences: astronomy, biology, physics, chemistry, geology, geography, anthropology, archeology, sociology, and so on. They each are characterized by their domains of inquiry and by their various methods and technological infrastructures. So, we need to speak of progress in terms of the individual sciences and their subfields. But even then, we run into the problems noted earlier, just finer grained. What does progress in astronomy look like? We know that astronomers are mapping more of the universe and producing more data regarding the number of galaxies and theories regarding dark matter. But it seems that the more we discover, the more problems we create. Just what is dark matter?

Finally, wouldn't finding the final big picture require that all those sciences in some way cohere or be integrated? That was the goal of the logical positivists who sought a universal language of science through which such a unification could be achieved, a goal that has been roundly refuted.

If there is no coherent account of a goal for scientific progress, it would then seem like we also can't measure that progress, for that would be to determine how much closer to the goal we have come. So, what is wrong with suggesting that we ought not to talk about the progress of science? What do we lose?

For one thing, we lose the propaganda value for science that comes with talking about progress. When we talk of the progress of science we make science sound like some rolling epistemological and metaphysical juggernaut. We often hear it said, "Science is our best effort at discovering what the world is like." That may be true, but talking about progress in science does not really add anything to that claim. In truth, if it can be shown that claiming there is progress in science is an incoherent claim, as I hope I have done earlier, then talk about progress actually diminishes the successes of science.

But if there is no progress in science, then how are we to evaluate new discoveries? This is where I must return to the notion of a coherent story and Wilfrid Sellars's account of the aim of philosophy. If, as Sellars claims, we aim to see how it all hangs together, then we need to understand what that means. To start with, it means that whatever picture we put together of the world and our place in it must not contain contradictions. To start there also entails a commitment to operating within a linguistic framework. Our picture of the universe is expressed in language. Stories are expressed in language. That much of physics uses mathematics does not nullify this point given the talk of the language of mathematics.

Further, a coherent story is one in which there are no loose ends. This means that until the end of inquiry, assuming that point comes along, there will always be loose ends. To assume a Peircean stance—inquiry is fueled by doubt, doubt that we have it right. And it is always possible to doubt a statement, any statement.

So, let us assume that we always have loose ends, that there is no end of inquiry, and that a fully coherent story is not possible. That does not mean that the idea of a fully coherent story can't serve as a regulative ideal, something we strive to produce. If our goal is a fully coherent story, then bits and pieces that don't fit either need to be rejected or pursued further. Take the example of dark matter/dark energy. We don't know what it is. Yet those who work in the area claim dark energy occupies 62% of the universe and dark matter approximately 38%. That is a pretty big loose end. If we don't know what it is or how it is supposed to behave, we are not in a very good position to build instruments or design experiments to detect it. Is it merely an artifact of the math of relativity theory, sort of like the minus version of an equation?

There is, in fact, an attempt underway to find dark matter. It involves, naturally, a highly refined piece of technology, an atomic clock. Starting in the late 1990s, Jun Ye, a physicist at JILA (Joint Institute for Laboratory Astrophysics), was determined to build the world's most accurate atomic clock.

> What Ye didn't know was that he was also designing a dark matter detector. That realization came in April 2015, when he got an email from Victor Flambaum, a physicist at the University of New South Wales in Sydney, Australia. Flambaum

> told Ye that according to certain theories, dark matter could subtly tweak the fundamental constants of nature, ever so slightly changing how fast clocks tick. Ye's latest device . . . was one of the few in the world sensitive enough to have a chance of picking up this faint signal. Ye is now [April 2018] carrying out one of the world's most unconventional searches for dark matter. The hypothesized substance is thought to outweigh "ordinary" matter 5 to 1, but so far it has made itself known only through gravitational influence. By carefully measuring whether the frequency of his clock meanders up and down in time, Ye could reveal an interaction between a hypothetical ultralight dark matter particle and the internal constituents of atoms. Such a find would shake up physics. ("Ultra-accurate Clocks Lead Search for New Laws of Physics," by Frederic Simon, *Quanta Magazine*, April 17, 2018)

In psychology, consciousness is another example of a loose end. The concept plays such an important part in our explanations of human behavior, yet we can't pin it down. We have yet to show how it is related to our neurophysiology and at the same time we are unwilling to give it the status of an epiphenomenon, something generated by our neurophysiology, but not itself physical, resurrecting something akin to Descartes' mental matter.

In human affairs, "love" is a significant concept, yet no one knows what it is. It is supposed to be something like a positive feeling one has toward another human being—but it also seems to be ineffable, and it is not even clear that it is like pornography, because you often don't know it when you see it because you don't know what it is supposed to look like.

Dark matter, consciousness, and love are all loose ends in our efforts to construct a coherent story about the universe and our role in it. And yet, we seem to be able to form a fairly understandable and workable picture of that world despite these loose ends in our current story. Human beings are really interesting creatures whose characteristics are not fully understood at this point, but we don't want to rule them out of our story, primarily because they are the ones telling the story!

In some ways, the problems we are wrestling with here are similar to the ones Sellars confronted when he divided the world into two images, manifest and scientific. Sellars thought that the picture of the world created by science and that of the world created by common sense had to be reconciled and the way to do that was to construct a stereoscopic image melding the two together. It didn't work because in the end he also insisted that "in the dimension of describing and explaining the world, science is the measure of all things, of what is that it is, and of what is not that it is not" (Science, Perception and Reality 173). This position entails an extreme scientific realism. This means that if we can't have a scientific theory of love, love does not exist—which is clearly nonsense, sayeth the poet. The problem stems from the bifurcation between the scientific and manifest images.

I prefer to put the problem differently. We humans are telling a story about how the world works and, perhaps most importantly, how we fit into that picture. To begin with, we need to recognize that we are not in a separate picture, but it is our picture from the start. Our goal is to explain the world around us, and from the start we must acknowledge that we are part of that world. What constitutes a good explanation is one in which what we know, see, experience, and so on hangs together, that is, coheres, in the manner we have been discussing. This story will change over time as the physical world we engineer changes and as our knowledge changes. This is not to give up on the idea of progress; rather it is to reduce it to something manageable.

There is yet another account of progress that few philosophers have discussed. One of those few is Nicholas Rescher in his *Scientific Progress* (1978). Rescher explores the ideas of a number of historians and scientists who equate scientific progress with the increase in the amount of science being done. Thus, if there are more resources being allocated to scientific research and more labs being funded and more students graduating and more publications, then there is progress. This emphasis on "more" leads him to explore what it takes to do more science and whether or not there are limitations to this kind of progress. Interestingly, he does not foresee an end to progress, but he does see whatever progress is being made slowing down. The primary cause of this slowdown is, as it turns out, economic.

The picture looks something like this. Discovering new significant things about the universe will become increasingly expensive since the cost of the technologies we will need to do the work will become more expensive. Yes, that is correct: as far back as 1978, Rescher understood that there was a connection between technology and science. His chapter 8 is titled "The Technological Dependency of Science." But he is, as usual, very careful to explain what he means:

> When one speaks of "*technology*" in the present context, the word must be construed appropriately, for what is relevant for present purposes is *the technology of data-generation and information acquisition and processing*—in short, the technology of scientific *inquiry* itself. (p. 134)

To this he appends a lovely footnote:

> The knowledge facilitating power of his instruments is apt to engender in the experimental/observational scientist an exultation in his instrument that is not unworthy of a lover. Thus Galileo: "O telescope, instrument of much knowledge, more precious than any sceptic! Is not he who holds thee in his hand made king and lord of the works of God"? (pp. 134–35)

Clearly, when Rescher speaks of the relation between technology and science he has a much more restricted conception of technology than I have been building up here. And when he addresses the question of the possible end of scientific inquiry, he focuses on the costs of making significant discoveries, where such costs eventually, he thinks, will be prohibitive. This does not mean that science will come to an end. It is rather that significant scientific discoveries of the kind that end up in major histories of science will probably slow down precipitously. The fault is not in our ability to understand, nor in the science, but it is in the cost of the necessary technologies. So, this account of progress, while it is an improvement over those accounts that concentrate on devising accounts of decision-making, ends up number crunching. But it is not clear to me that more of something is necessarily better, and there is an undertone of "better" when talking about progress.

If we cease talking about scientific progress and concentrate on the individual sciences, maybe we can make some headway. Taking this route means looking at how each of the sciences defines its goals and methods and problems. Thus, while there may be no scientific progress *tout court*, there can be problems solved.[3] It is also the case that by concentrating on the problem-solving activity in the individual sciences, we get a better feel for the relationships between the theories that guide their research activities and their technological infrastructures.

An example of the sort of thing I have been discussing concerns efforts to find the mass of neutrinos. KATRIN is the anagram for the Karlsruhe Tritium Neutrino experiment, a 60 million euro project designed to do just that. The neutrino appeared on the scene in 1914, when James Chadwick was studying beta decay. Beta decay is a

> form of radioactive decay in which a nucleus emits an electron, transforming a neutron into a proton. Conservation of energy suggested that the electrons from a particular nucleus, say lead-214, should always emerge with the same energy. Instead, Chadwick showed that they emerge with a range of energies extending down to zero, as if energy were disappearing. (Cho, Adrian, 30 June, 2017, pp. 1322–1326)

KATRIN measures beta decay in tritium. Enrico Fermi was working on a theory of beta decay in the 1950s. Fermi

> realized that the electrons' energy spectrum holds a clue to a key property of the neutrino: its mass. If the particle is mass-less, the spectrum should extend up to the same energy the electron would have if it emerged alone—corresponding to decays in which the neutrino emerges with virtually no energy. If the neutrino has mass, the spectrum should fall short of the limit by an amount equal to the mass. To weigh the neutrino, physicists had only to precisely map the upper end of the electron spectrum in beta decay. (Cho, Adrian, 30 June, 2017, pp. 1322–1326)

It sounds very simple to do, but it is not. The experimental layout

> depends on a device called a windowless gaseous tritium source: an open-ended pipe 10 meters long that tritium enters from a port in the middle. Superconducting magnets surrounding the pipe generate a field 70,000 times as strong as Earth's. Beta decay electrons from the tritium spiral in the magnetic field to the pipe's ends, where pumps suck out the uncharged tritium molecules. Set it up right, with not so much tritium that the gas itself slows the electrons, and the source should produce 100 billion electrons per second.
>
> Finally, physicists must measure the electrons' energies. That's where KATRIN's zeppelinlike vacuum chamber [comes] into play. Still riding the magnetic field lines from the source, the electrons enter the chamber from one end. The magnetic field, now supplied by graceful hoops of wire encircling the blimp, weakens to a mere six times Earth's field as the field lines spread out. The spreading is key, as it forces the electrons to move along the lines, and not around them.
>
> Once the electrons are moving in precisely the same direction, physicists can measure their energies. Electrons lining the chamber create an electron field that pushes against the onrushing electrons and opposes their motion. Only those electrons that have enough energy can push past the electric field and reach the detector at the far end of the chamber. So, by varying the strength of the electric field and counting the electrons that hit the detector, physicists can trace the spectrum. KATRIN researchers will concentrate on the spectrum's upper end, the all-important region mapped out by just one every 5 trillion electrons from the decays. (Cho, 2017, p. 1325)

The zeppelin-like vacuum chamber is the size of a blue whale. Lots of problems came up, and the experiment, now set to go, is nine years behind schedule. The whole setup is enormous, incredibly complicated, and very expensive.

In a detailed discussion of the project in Cho, as above, 2017, pp. 1322–1326, Adrian Cho wrote,

> KATRIN physicists have no guarantee that they'll succeed. From very different kinds of experiments—such as giant subterranean detectors that spot neutrinos from space—they now know that the neutrino cannot be massless. But in recent years, data from even further afield—maps of the cosmos on the grandest scales—suggest that the neutrino might be too light for KATRIN to grasp. Still, even cosmologists say the experiment is worth doing. If the neutrino mass does elude KATRIN, their current understanding of the cosmos will have passed the test.
>
> A definitive measurement, on the other hand, would be potentially revolutionary. "If KATRIN finds something," says Licia Verde, a cosmologist at the University of Barcelona in Spain, "cosmologists will be left scratching their heads and saying 'Where did we go wrong?' "

This reveals several interesting things. First KATRIN is a very good example of Big Science. Second, cosmologists are doing an experiment that, if it finds the mass of a neutrino, will falsify current theories. Third, their hunt is based on a lot of other Big Science projects. In a word, the technological infrastructure of this science is being used to solve a problem. If it finds that neutrinos have weight, a new theory will be needed. If that happens, it will be a perfect example of the way in which technological innovation forces theory change as this experiment involves a novel experimental structure. But it is not the case that forcing such a change is progress in any meaningful way.

There is some indication that the complicated relations between technologies and the sciences are being recognized. The August 31, 2018, issue of *Science* was a special issue entitled "Technologies Transforming Biology," although in the editorial by Editor-in-Chief Jeremy Berg, we find the claim "New technology is one of the most powerful drivers of scientific progress." More heartening is the comment on the picture on the cover:

> A Mobius strip symbolizes the infinite cycle of technology and science. Researchers have embraced revolutionary biotechnologies—such as super-resolution microscopes and CRISPR-based genetic engineering—to advance scientific frontiers, and the knowledge gained can often be used to develop even more powerful technologies. (Berg. Jeremy, 2018, p. 827)

One explanation of the attempts to talk about scientific progress is the extent to which we tie it to the idea of human progress. We really don't know what that means either, but we have managed to make it part of our narrative. We can point to certain benchmarks like eliminating certain diseases. And we can cloak the whole hazy notion with the idea of making human life better. But when faced with the facts, that seems like a hopeless goal. We seem incapable of stopping the out-of-control explosion of the number of people on this planet. Instead of making sure everyone is well fed and housed, we find famine and collapsing cities and shrinking resources, possibly leading to more conflicts. The claims for human progress come down to a few specifics that support the illusion of progress by focusing on shiny baubles like satellites, travel to the moon and Mars, genetic detection of potential health hazards, and so on. In effect, talk of human progress boils down to technological progress. But what constitutes technological progress?

NOTES

1. Clearly, the positivists saw progress as a characteristic of science—heading toward a unified science. But no one is a positivist anymore, so maybe there is progress somewhere.

2. This clearly is not going to sit well with social constructivists. But the fact of the matter is that the universe is not the product of what we collectively think.

3. For a detailed analysis of problem-solving in the sciences, see Larry Laudan, *Progress and Its Problems*.

Chapter 8

Technological Progress?

In an earlier work (Pitt 1999), I argued that there is no such thing as Technology with a capital "T"—there are various technologies: automobiles, computers, toasters, and so on. But faced with the need to come up with a definition of "Technology" to satisfy my editor, I proposed "humanity at work" and went on to argue that the definition was so all encompassing as to be useless in that form. But taking that definition seriously works if we look at specific technologies and what people do with them, a farmer and her hoe, a farmer and her tractor, a farmer and her drone. The farmer uses the hoe to work the earth in small spaces. She uses the tractor when dealing with large spaces like mowing pastures, or harvesting hay. She uses her drone to check on the health and status of her crops. But all of this together does not constitute Technology with a capital T (TWACT).

The search for a definition of TWACT is the search for some kind of essence of Technology. This was Heidegger's project in his "The Question Concerning Technology" (1954). His answer involved drawing several distinctions. There is a difference, he notes, between being correct and being true. There is also a difference between giving a definition and finding the essence of something. Finally, there is a difference between ancient technology and modern technology. What this comes to is that modern technology has the capacity to be used against humanity and, therefore, is evil. Does that mean that the essence of modern technology is evil? That is unclear. But what does seem clear is that we have cast little light on the subject by going down the road of defining essences.

OK, so if there is no such thing as TWACT, what sense can we make of the notion of technological progress? You can see it coming: well, if there is to be progress, then there must be some goal TWACT is striving to achieve—note that Technology is not doing anything, as we saw in chapter 6; people are doing the doing or not doing the doing. So, let's start with the obvious. In general, everyday discussion, "technological progress" refers to people using ever more "advanced" technologies to make their way around in the world. There are two problems here: (1) Does this commit us to some form of technological determinism? (2) What makes a technology more advanced? Let's deal with the second question first.

Consider the case of the telephone. Up until 20 years or so ago, when you talked about a phone you meant this clunky device that was tied to a physical line. Today, landline phones are almost an anachronism. The question is, then, is the mobile smartphone more advanced than the landline phone? I want to argue that it is a different device, not more advanced. It allows the user to do more than merely communicate with another person by voice over a long distance. It does lots more; among other things it allows you to access the internet, social media, text while driving, play games, search the web, get weather forecasts, keep going. For most of the life of the landline telephone, there was no internet, so in a very specific sense the mobile phone cannot be an improvement since the thing it is improving could not use the kind of thing that makes the mobile phone so versatile, that is, the internet. So, it is the internet, among other things, that made the mobile phone possible. The landline phone is still around and is still being used; I have one in my home and another in my office.

Let's try another device, the automobile. I recently purchased a new vehicle that came equipped with improved pollution controls, gets better mileage, and requires less maintenance. It also came with a slew of electronic stuffs that were not previously available—GPS, warnings of traffic problems, access to satellite radio, telephone, temperature control, and more, so much more I have not had the time, even after two years, to figure it all out. Is this vehicle "more advanced" than my old 2001 car? Again, I am inclined to say it is a different device—not just a means for getting from point A to point B.

Again, the internet is not an improvement on anything—it is something different. What this all comes down to is an admission that I don't know what it means for something to be more advanced than something else. In what sense are contemporary *Homo sapiens* more advanced than Neanderthals? In what sense are today's intercontinental missiles more advanced than World War 2 V2s? Yes, they can travel further. But that doesn't make it more advanced. It can carry a heavier payload—but is that an advancement? No obvious answer jumps out.

But is it true that technological progress means devices that make it easier or more efficient for people to make their way around? If so, how do we accommodate the possibility that technological progress may eliminate work? (So much for my definition: If technology is humanity at work and eventually there is no work for humanity, does that mean there are no technologies because the technologies are doing the work, not people? Or does that mean the technologies have become human?)

I am referring to the development of such technologies as autonomous vehicles, artificial intelligence, self-programming machines, and so on. As our technologies become increasingly capable of performing the tasks we are currently burdened with, what will there be left for humans to do? And how will that affect our sense of who and what we are?

It is well known that work is important to one's sense of self-worth. Contributing something to improve the common good is essential to a sense of well-being. It is possible that once relieved of certain burdens we might discover other things to do—but would they be as fulfilling as what many of us do now? It could be argued that much of what we do now is sheer drudgery and that the alleged sense of fulfillment is more a function of a cultural construct that substitutes work for God. Freed of the drudgery, people could begin to explore their natural talents to develop a new and exciting world of creativity. I am not sure what that actually means, but it suggests that each of us has hidden creative capacities. Whether or not that is true remains to be determined, but I am constantly surprised at what my students come up with when given a free hand.

The problem we are addressing here, technological progress, has serious ramifications for how science is done. I will return to this issue after taking up the threat of technological determinism.

When first addressing the question of technological change discussed earlier, I noted two questions had to be addressed. The first concerned technological progress, and we have seen somewhat where that takes us. The second was technological determinism, a hot topic in philosophy of technology for many decades. The question here is, does technology have a kind of momentum that necessarily drives its development in certain directions? For example, once the Germans developed the V2 missile, were we inexorably set on the road to space exploration?

The Technological Determinacy thesis (TD) admits of many versions, weak, strong, sorta, kinda, and so on, and they have been treated at length by others. I will not rework worn-out material here. Instead, I want to explore the tie between TD and progress.

The question is, are we fated to make progress? Is there something in the human makeup that pushes us to make better things, do more, be more creative? This is a kind of determinism that seems to make sense when we look at what we have done already. After exploring our world, we are now seeking others to explore, or, to borrow from Star Trek, "to boldly go where no man has gone before." But isn't that part of the same myth that sees human progress in terms of making things better? Or even a kind of manifest destiny?

If we are to be honest with ourselves, a different perspective on what we as a species have done yields not so encouraging a view. Not attempting to be controversial, but it seems the frantic activity that many see as progress is really a headlong drive toward oblivion. As noted earlier, we are rapidly overpopulating the planet. The ramifications are enormous. Couple that with climate change and the overall destruction of the environment, the use of fertilizers that create dead zones in the oceans, and so on, and we have full-scale disaster pending. Further, we have failed to stem the spread of nuclear weapons. The possibility of nuclear conflict is real. The prospect of a decimated planet with life (possibly) beginning over is real. The prospect of the end of humanity brought about by the technological "progress" of humanity is real.

However, it isn't technological progress that is the culprit, but it is technological *development*. As I have been arguing, there is no clear goal that we can point to against which we may measure our progress. But there is a lot of technological development and without sounding like a Luddite, it needs to be checked.

The out-of-control technological developments we see around us provide us with the illusion of progress, but only if we use rose-colored glasses. Nature is being trampled. We leave our garbage as our footprint as much as something helpful. We operate on the principle that if we can do it, let's do it. But that is the wrong principle. Rather, we ought to be asking ourselves, "If we can do it, should we?"

Clearly, the doomsday scenario inadequately sketched earlier needs a lot more to constitute an argument. But the basic point can stand. We should not be talking about technological determinism or technological progress, but rather technological development because it is the development of newer, bigger, more complicated, and invasive technologies that is creating many of the problems that we are facing and that threaten our future. We also need to develop the means by which to assess that development and propose further developments. But this becomes increasingly difficult as we continue to rely on the technologies we have created and as we ignore the *fundamental law of technological development: unintended consequences*. The point here is that in order to figure out what the impact of a given technological development will be we have to know what other technological developments are taking place at the same time or are ongoing as well as what else is being proposed.

Consider the following example. As I write, the United States is in the midst of a pipeline building frenzy. These pipelines carry oil, natural gas, fracked gas, and similar materials. One such 330-mile-long pipeline, the Mountain Valley Pipeline (MVP), is being built from the Marcellus Shale in northeastern West Virginia, south through several West Virginia counties before crossing into Virginia heading east to hook up with a north/south pipeline leading to export terminals. I will avoid the politics of the situation as much as possible and concentrate on one unforeseen consequence. As it crosses into Virginia, the first county MVP will cross is Giles County (this is the county in which I live). This is a poor mostly agricultural county. It is also the most seismically active county on the East Coast of the United States. It is very rugged terrain in the Appalachian Mountains. Not really suited to industrialization, the county has taken advantage of its natural beauty and resources to create an eco-economy, an economy built on attracting tourists interested in taking advantage of its rivers (the New River is the second-oldest river in the world) and hiking trails, which include a portion of the Appalachian Trail.

Enter MVP. The proposed pipeline will be buried 10 feet down. To construct it, forest, some of it old growth, will have to be cleared since the construction plan calls for a 150-foot-wide construction corridor. When the pipeline is completed, a 50-foot-wide clear zone will have to be maintained, leaving an ugly scar across the face of the county, spoiling beautiful views and destroying the tourist economy. Restaurants and other eateries will be empty, hotels empty, meaning that local people will be out of work.

The unintended consequences of building this pipeline are a wrecked economy and the effects that will have on the lives of the local population. In the planning for this project, no one checked to determine the economic impact it would have. The federal agency responsible for granting permission to build pipelines, the Federal Energy Regulatory Commission (FERC), did not ask for that information. Jobs are scarce in southwestern Appalachia.

The key issue is not technological progress or technological determinacy but *how to assess technological development*. It seems clear from our history that no technological project is without unintended consequences. It also seems clear that the difficulty of calculating consequences is compounded by the fact that multiple projects are always simultaneously underway, in this case, the building of a pipeline and the development of an ecology-based economy. Assessing how they impact society and the environment cannot be accomplished successfully without including how they impact each other and what the unintended consequences of all those interactions might be. How to do this is difficult due to the simple fact that technological projects cannot be thought through to the smallest detail ahead of time, and, as we saw with the Hubble, sometimes it is the smallest detail that derails the project or calls for a lot of correction.

To return to the pipeline example, at present there is no way to predict seismic activity accurately. Thus, there is no way to ensure a pipeline built in karst geology in a seismically active zone will not rupture, or worse, explode, doing even further damage to the eco-economy. In the immortal words of Majorie Grene, “It’s all very complicated.”

It might be possible to do some serious forward planning using the enormous computer power we now have available. All we need to do is list all the possible relevant variables and write some code that projects possible outcomes and their likelihood of developing. Sounds rather straightforward, doesn't it? Now consider the details. If we consider just two projects, the MVP and the eco-economy of Giles County, we will see that this is not as straightforward an approach as presumed. By the way, this is not a case study—it is an example of how hard it is to predict the consequences of unchecked technological development.

To begin with, we need to consider the effect of digging a 10-foot-deep trench through Karst terrain going over and down steep mountainsides that have a long record of avalanches. The Karst terrain is porous limestone in which there are numerous underground caves and sinkholes. The effects of severe weather on this terrain are unpredictable, and the weather itself is increasingly unpredictable given climate change. The pipeline is built in sections and welded together. The welds are the danger spots since leaks occur at those joints, and if the ground moves, as in an earthquake, those welds are placed under severe stress and could break, leading to the possibility of an explosion. The blast zone, should it explode, is approximately ¼ mile on each side. These are heavily forested mountains, so the possibility of large fires is real, but how to calculate it? We can't predict earthquakes, yet. We have difficulty predicting the weather, let alone earthquakes. These factors will certainly affect the eco-economy, but it is hard to say how. Finally, we have no idea what impact all of this drilling and blasting will have on the water table. Most of the county gets its water from wells and streams. If it were merely a matter of providing water for humans, it might be possible to manage any destruction to the water table. But this is an agricultural area, and livestock take a lot of water.

A reasonable response is to go ahead and model the problem, develop the computer programs, and tweak them as we find out more. The major problem here is trying to figure out the relevant variables. If we return to our pipeline example, we can get a sense of the difficulties. We have already raised the issue of earthquakes and karst terrain. We have yet to figure out how to predict earthquakes. We also have not figured how to model the effects of an earthquake on the karst terrain. We know there are lots of caves and sinkholes in this county. How does an earthquake affect them, and what is the resulting underground topology look like and how do we adjust to the new configuration as we attempt to predict what will happen with the next earthquake? I am trying to paint a picture of an unstable environment and the problem of modeling the effects of an earthquake on it and trying to predict the impact of such a quake on a pipeline buried 10 feet down. These are just two variables, but they are complicated variables and, I suspect, this situation is not uncommon. What I mean is that there are situations that are very complicated, even if the general description presents them as fairly simple.

Predicting and developing the means to handle unintended consequences in these situations begins to look impossible.

For example, while it may look fairly stable from 30,000 feet up, the world economy is a delicately balanced set of tugs and pulls. Mess with one little part, and the ramifications can be catastrophic. Let us assume that one country, the United States, decides to place tariffs on a certain set of goods, steel and aluminum, that are imported from several other countries, the EU, China, Australia, India, Russia, and Canada. In retaliation, those other countries place tariffs on targeted U.S. goods, motorcycles, bourbon. The American motorcycle company decides to open a plant in Europe to save costs to European buyers. The initial policy was intended to strengthen the American economy. Now we have American companies moving to Europe, costing American jobs, and we have yet to see how the other pieces will begin to fall. One other set of consequences are the political costs, as many of the targeted countries are political and military allies. How is the imposition of economic penalties going to affect those alliances? Now, one ought to have thought that this scenario could have been foreseen and planned for. But it wasn't. How do we plan for the consequences of every policy decision? But that is what we have to do, if we are to avoid unintended consequences. Close a factory and hundreds of peoples' lives are affected.

The point I have been trying to develop here is that actions do not take place in isolation. They take place in a complicated context of many other actions and situations. Technological developments do not take place in isolation. To determine if you ought to do what you are thinking about doing, you must consider what else is being done in that context as well. Yes, I railed against "context" in chapter 1, but what I am referring to here is not some abstract thing, but the sum total of circumstances and other decisions that are affected by and affect the results of your actions. To plan well for technological development, one must factor all these in.

Chapter 9

Technological Development and Scientific Change

We now turn again to the impact of technological development on science, but it is clearly too large a topic to cover it all. We can gain entry by picking another example to make the point about the impact of technological innovation: artificial intelligence (AI).

We are looking at a major change coming soon to a lab near you. One of the biggest growth areas in technology realms is AI. The July 7, 2017, issue of *Science* was devoted to an examination of the impact of AI on the doing of science from astronomy to particle physics to microbiology. AI, sometimes referred to as machine learning, is making serious developmental strides daily. We are talking about sophisticated computer programs that can teach themselves to recognize patterns, carry on a conversation, do research, enhance our ability to make sense of large data, and so on.

There is concern in some circles that AI will be putting people out of work. However, while it is being used in industry, and while some jobs are being appropriated, other jobs are being created. While it is too soon to be able to draw any concrete conclusions, it does not seem like the dystopia predicted by Luddites will come to pass. However, in the lab, while we are seeing very positive results, we are also witnessing the displacement of people. Graduate students are one of the first to feel the pinch. According to John Bohannon in "The Cyberscientist" (Bohannon, John, 2017, pp. 18–21), robots, physical things guided by AI, are taking over the labs. Bohannon gives an example of robots used by Zymergen, a biotechnology company outside of San Francisco. The robots

> spend their days carrying out experiments on microbes, searching for ways to increase the production of useful chemicals. Here's one called Echo. Nestled within a blocky jumble of equipment, a robotic arm grabs a plastic block

> dimpled with hundreds of tiny wells carrying liquid. A laser scans a barcode on the block's side before Echo loads it into a tray. What happens next is too subtle for the human eye to perceive. . . . Instead of using a pipette to suck up and squirt micro-liters of liquid into each well—a tidal wave of volume on the cellular scale—the robot never touches it. Instead, 500 times per second, a pulse of sound waves causes the liquid itself to ripple and launch a droplet a thousand times smaller than one a human can transfer.

None of the technology, the robot, the barcode, and so on, is new. What is new is the fact that a computer program designed the experiment. The scientist in charge does not know what the experiment is supposed to do.

> "I want to be very clear," says Zymergen CEO Joshua Hoffman, heading off a persistent misunderstanding. "There is a human scientist in the loop, looking at the results and reality checking them." But for interpreting data, generating hypotheses, and planning experiments, he says, the ultimate goal is "to get rid of human intuition." (p. 19)

Why? First of all, the kind of manual labor required for the experiments is tedious. Second,

> A graduate student's tiniest mistake or imprecision can waste weeks of work. Even worse is a sloppy decision by the postdoc who designed the experiment for that student, wasting months of effort. (p. 19)

In short, AI, or machine learning systems, coupled with a technological infrastructure of robots, can take the drudgery out of much experimental work in the biological sciences and produce more precise results. But graduate students, future scientists, learn their craft doing much of the drudgery. Furthermore, the cost of their graduate education is usually paid for through a grant their major professor has procured. Replacing graduate students with AI must eventually result in rethinking how science graduate students are paid, what kinds of work they will do, and what kind of education they will need and receive.

However, if science is what scientists do, and if more and more of the "science" is being done by machine learning systems, then is it really science? That is where our newly developed understanding of what is required to do science comes in. You can't "do science" in isolation. It requires a technological infrastructure. Machine learning systems have changed the technological infrastructure of science. This is a new technological development that enables more precision in experimental work. This is an example of how technological developments change the way science is done and what we can do and learn.

If we turn to astronomy, the results are equally impressive. Astronomers work with pictures taken by a variety of types of telescopes: optical infrared, X-ray, and so on. The images they obtain are rarely nice and sharp. More often than not, they are hazy, blurry, and full of noise. The solutions to crisping up those images boil down to two. (1) Build better telescopes and (2) find a way to sharpen those images and eliminate the noise. Telescopes are frightfully expensive and take decades to design and build. Using AI to sharpen those images is already here and producing amazing results, and it is a lot cheaper.

In that same issue of *Science*, there is an article reporting on the results of Kevin Schawinski, who developed a generating adversarial network,

> A kind of machine learning model that pits two dueling neural networks against each other. One is a generator that concocts images, the other a discriminator that tries to spot any flaws that would give away the manipulation, forcing the generator to get better. (Sokol 2017, p. 26)

Schawinski used this method to generate four pictures and then posted those four pictures of fuzzy galaxies and asked his fellow astronomers to classify them. Someone suspected fraud and asked if they were real galaxies.

> At ETH Zurich in Switzerland, Schawinski, computer scientist Ce Zhang, and other collaborators cooked up the galaxies inside a neural network that doesn't know anything about physics, it just seems to know how galaxies should look. (ibid. Sokol)

The objective of developing this technique was to extract from the millions of pictures of the heavens more information by sharpening up fuzzy images.

> Schawinski's team took thousands of images of galaxies and then artificially degraded them. Then the researchers taught the generator to spruce up the images again so they could slip past the discriminator. Eventually the network could outperform other techniques for smoothing out noisy pictures of galaxies. (ibid. Sokol)

The new technology, machine learning using neural networks, has made it possible to use data produced by many telescopes in ways that give us usable information to extend our knowledge of the universe. The generating adversarial network has significantly modified the technological infrastructure of science. At the heart of machine learning is the neural network. What is it, and how does it learn?

A neural network is a computer program loosely organized in the manner of the human brain. It "learns" by being fed examples of items that it then organizes by the structural principles with which it has been programmed. In that same issue of *Science*, Paul Voosen explains it this way:

> At one end sits a messy soup of data, say millions of pictures of dogs. Those data are sucked into a network with a dozen or more computational layers, in which neuron-like connections "fire" in response to features of the input data. Each layer reacts to progressively more abstract features, allowing the final layer to distinguish, say, terrier from dachshund.
>
> At first the system will botch the job. But each result is compared with labeled pictures of dogs. In a process called back-propagation, the outcome is sent backward through the network, enabling it to reweight the triggers for each neuron. The process repeats millions of times until the network learns—somehow—to make fine distinctions among breeds. (Voosen, 2017, p. 24)

Interestingly, the people creating the neural network don't really know how it learns and as the systems get more complicated, the black box nature of neural networks will increase. The problem for us then becomes knowing when to trust their results.

A colleague of mine, Ben Jantzen, is working on ways to facilitate discovery in science using some fancy algorithms. According to Jantzen:

> Our algorithms work with raw data—sometimes small, sometimes large—and provide a measure of the similarity of the dynamics of two or more systems of interest (e.g., ecosystems, metabolic processes, city power grids, etc.). What makes them useful is that they can indicate similarity or dissimilarity even when *nothing* is known about the dynamics, or even whether we've chosen an appropriate set of variables in terms of which to describe the dynamics. These tests can be turned around different ways to accomplish tasks like:
>
> —determine which systems are governed by a shared scientific law, even if that law is unknown
> —determine when a system's dynamics has changed (e.g., when the structure of an ecosystem has been altered)
> —validate a model or simulation when closed-form laws are unavailable for both the simulate and the target system
>
> If I had to put it in one sentence, I'd say the basic idea is that our algorithms can tell you when two things are governed by the same regularity, even when we lack the ability to explicitly identify that regularity. (Personal communication, November 3, 2019)

AI is clearly transforming the way many activities in science are conducted. The more complicated the tasks, and the more inscrutable the processes, the question arises as to why we should call this science. The reason for that is that in our hopeless search for a definition of "science" we failed to find a useful general accounting that covers all the various activities we associate with science. We are left, then, with something like "science is what scientists do." But if scientists are increasingly removed from the day-to-day activities, what do they do? And we face again the question of what counts as science.

Perhaps that is the wrong question. Maybe we should be trying to answer a different question: "Who counts as a scientist and why?" Here we have two possible paths to take. We could do the typical philosophy thing and try to develop an abstract definition, or we could do some empirical work. Given the tenor of my previous remarks, it should come as no surprise that I opt for the latter.

Let's stick with universities. The usual advertisement for a faculty member will state something like: must have a PhD in, for example, astrophysics. So, at least at the university level, an advanced degree is one of the conditions for being hired into a science department. Wait a minute, we have jumped too soon into the quicksand. Perhaps the prior question is, what are the grounds on which a department is classified as a science department?

Before the French introduced the term "scienza"—the domain of knowledge fell in two parts, natural philosophy and moral philosophy. What we call the sciences today fell out of natural philosophy. It is not clear that there were science departments until the invention of the research university in Germany in the 19th century.

To get a sense of how arbitrary the identification of a field of study and research is, consider what happened in my university. The College of Arts and Sciences at Virginia Tech was dissolved and two colleges were formed: the College of Science and the College of Liberal Arts and Human Sciences. The grounds on which a department joined the College of Science were both obvious and not. Certain departments were clear candidates: physics, chemistry, biology. Other departments that wanted to join the College of Science could, or so it seemed. Psychology and economics opted in. So did mathematics and statistics. Geography went to the College of Agriculture and Life Sciences. I was head of the Department of Philosophy at the time and half in jest suggested to the dean of the College of Science that philosophy join his college. We had and still have a substantial number of faculty members whose specialty is the history and philosophy of science, so it was not a totally outrageous idea. He said, "Why not?" Needless to say, not all of the philosophy faculty thought that was such a good idea. But the fact that certain departments seemed to be obvious choices doesn't answer the question of what are the criteria for being deemed a science and thereby a good fit for a College of Science.

So back to the question of why were physics, chemistry, and biology obvious choices? One possible consideration is historical. These departments have traditionally been considered sciences. But why? What is it that goes on in these departments that qualifies them to be called sciences? As noted earlier, the term "science" derives from the French "Scienza" for "knowledge." OK, but other fields produce or investigate knowledge; consider management, child development, philosophy. Oh, oh, I sense we are heading to the definition game and frankly I don't want to go there.

Earlier I claimed that the mark of knowledge is successful action. If you study music performance and learn to play the piano, that counts. If you are a member of the music faculty and a composer and produce works that people want to play—that counts.

It looks like what counts as a science is a matter of historical contingency. So, let us return to the question of the criteria for being a scientist of one kind or another. We started out by noting that having an advanced degree in the appropriate area was a start. It also seems to be the case that having that degree awarded by a PhD granting institution is another criterion. Now it is always possible for these criteria to be sidestepped. Let's say you have a PhD in philosophy but no science training. On the other hand, you have published a number of papers in serious science journals. That might be enough to let you slide by. But we may have also stumbled on another criterion: publication in appropriate journals. But there is a problem. Several of the faculty in my departments have adjunct appointments in other departments: geography, computer science, and economics. Does that make them geographers, computer scientists, or economists? Well, the person with the adjunct appointment in computer science is supervising a computer science MS student's thesis.

Let's face it, the criteria are place-specific and flow with the times. In short, we don't know what makes someone a scientist of a certain sort for sure. But does it matter? What seems to matter is whether or not the individual in question can produce research that is accepted by others who count by the same slippery criteria.

Returning to the discussion of AI transforming science, our foray into an attempt to develop criteria may be irrelevant since AIs are taking over the jobs that scientists have done. That may mean that scientists are going to have to find different things to do, which is perfectly consistent with the changing nature of everything.

Chapter 10

Technological Infrastructures and the Social

At this point, I would like to expand on the ideas floated earlier in chapter 9 and look at some of the technological infrastructures that frame our ordinary lives. In the past, I have called them social technologies (Pitt 2000). But that encourages us to think of them as isolated systems when, in fact, no system is isolated. In looking at the social world, I also have the opportunity to return to my first love: the law. Education will also be discussed, as will the food system. What I am aiming to show is the incredible complexity of our contemporary society and how viewing it in terms of technological infrastructures helps to expose that complexity.

The law is a technological infrastructure designed to bring order to society. It enables society—it makes it possible. It has multiple components ranging from an international system to local policemen. There are legislatures: international, national, state (I am using the U.S. as my model here), county, and city. They are responsible for enacting the laws, regulations, and rules that are supposed to structure social relations. There are police forces. There are courts. Where there are courts, there are judges, clerks, lawyers, investigators, court buildings, and jails. But there are also different kinds of courts, from the local traffic court to state courts to state appeal courts to state supreme courts. The relation among these various courts is hierarchical: a conviction in a traffic court can be appealed to the court at the next level and then the next and the next. The ruling of the court is called a decision. Decisions at each level can be appealed all the way up to the U.S. Supreme Court, which has the final say. At the federal level, there are district courts, appeals courts, and the Supreme Court. Decisions are often cited as setting precedence, which constitutes a kind of justification for the decision the judge or panel of judges arrive at. The kinds of things that can make their way through the system range from simple traffic convictions or jail sentences to challenges to the laws themselves.

This is sketchy, but it gives us enough to work with. Let's start with the lawyers. To qualify as a lawyer, someone who you go to to represent you on the assumption that she knows a lot more about the legal system and ways to maneuver through it than you do, the lawyer must be certified. She does this by passing the bar exam for the state(s) in which she is practicing. The bar exam is prepared, administered, and graded by representatives of state bar, usually lawyers. She prepares for the taking the bar exam in one of two ways: going to law school or reading for the law (in states where that is permitted). Law schools are colleges staffed by faculty who have studied the law or even practiced it themselves. Some law faculty have advanced degrees beyond the BA and JD, some may not have practiced. Some may not have a JD; for example, someone teaching jurisprudence may have a PhD in philosophy. Very few law schools stand alone—they are generally part of a university with its own complex structure of colleges, usually a baroque administrative structure requiring specialized knowledge, a campus—classroom buildings and labs, support staff, groundskeepers, faculty and students, athletic facilities and activities, alumni and alumni organizations, and fund-raisers.

So far, we have described a somewhat-complicated system, but if we look deeper there is more involved. To begin with there are two different sets of lawyers, public and private. Private lawyers are generally members of a firm, possibly their own. A firm is a business. It needs staff, secretaries, legal aids, accountants. Each of these needs special training, which usually means special educations—which means a tie-in to different educational institutions, business schools, and community colleges. Some community colleges are private, but most are part of a state-run community college system with its own rules and regulations and financing—which means a different relationship to the state legislature. Business schools are usually located in a university. Public lawyers work for various governmental agencies such as the office of the attorney general. They usually are the prosecutors, if the case involves a broken law of some sort.

In general, there is no special training for judges. They are usually recommended by a special committee of the state bar to a special committee of the state legislature, which then decides whether to bring the nominee's name to the full legislature. The bar consists of all the lawyers in the state who have passed the bar and paid the appropriate dues. It is structured in ways that permit it to certify lawyers and to recommend candidates for judicial appointments within the state court system, among other things. In other words, there are committees for this and committees for that, and so on.

We have looked at some of the various components of the legal system, but there is still more. Laws are legislation passed by legislatures and signed by a chief executive at the appropriate level. Laws are usually not passed at the city and county levels, but rules, regulations, and ordinances are. U.S. legislatures are bicameral. Pieces of proposed legislation usually originate in some committee or other charged with regulating that area. To say that politics is involved is to understate the problems. Usually, legislation originates in a committee in one house or another and then, after it passes, proceeds through a similar process in the other chamber, maybe with some negotiation between the two houses before the final legislation is sent to the executive. But there is more involved here than just some legislators introducing and arguing over proposed legislation. Each legislator has a staff. The composition of the staff varies, but there will at least be a legislative assistant (who writes the proposed legislation coming out of that legislator's office), a secretary, a liaison, and, depending on the rank and importance of the legislator, others. Likewise, the running of each house of the legislature requires a staff serving various duties and requiring different training.

If we turn now to the police, we have a different, but somewhat similar array of duties and structures. There are usually three levels of police in any given state: local town police, county police or sheriffs, and finally state police. Sometimes they work together and sometimes not. But there is another group of police they interact with: federal police. Federal police come in several varieties. There is the FBI and members of the Transportation Security Agency. Then there are the members of the state National Guard. Each of these different police men and women requires different kinds of training, which, in turn, requires they attend different types of educational institutions or events, which then have their own staffing needs.

As we have seen, the members of the legal profession have different and varying educational needs. But they are not the only ones who interact with the educational system. As a matter of fact, the educational system serves all of society, as does the legal system. That system is also complex. It consists of kindergartens, elementary schools, middle schools, high schools (these first four are often referred to collectively as K–12), community colleges, four-year public colleges, four-year private colleges, public universities, private universities, and medical schools which may or may not be affiliated with universities. Staffing these various institutions are teachers, administrators, librarians, support staff, bus drivers, and coaches, who at the K–12 level may also be classroom teachers. K–12 teachers generally need special training, and that usually takes place in schools or colleges of education generally associated with universities. In middle and high schools, there is also a lot of emphasis on sports, music activities and bands, and choirs. As already noted, the coaches are often also teaching faculty who got their coaching training usually by playing a collegiate sport. Community college teachers usually need a master's degree in the subject they will be teaching. Four-year colleges and universities usually require a terminal degree in the field of the faculty member. Four-year colleges usually place a high premium on teaching, but it is rarely the case that the faculty are given any formal teaching training. Universities usually place a high premium on research, grants, and training graduate students. Again, little or no training in these activities other than what one absorbs as a graduate student. Faculty are supposed to absorb these abilities by osmosis from their graduate mentors apparently. The administration of all of these institutions requires a variety of individuals with differing talents. The tracking of student records and determination of their eligibility to graduate requires a lot of time. Likewise, at the collegiate level, student financial aid requires special expertise. At research universities there are demands for assistance in getting grant proposals to the appropriate places; managing the grants once they come in; fostering research groups; and connecting with the wide variety of funding sources, private and public. Librarians at universities are increasingly computer literate as so much material is now being made available online. Driving a school bus requires special training, taking us into the world of automotive training.

In addition, in K–12, there is a lot of interaction with the community, especially with parents and the community source of funding. In the United States funding K–12 is usually a local affair. Generally, the source of those funds is property taxes, with the states rarely supplementing those funds by very much. That means community governments such as the county board of supervisors and the school board are in close and constant interaction with the members of the community. For public four-year colleges and universities, there may be some contact with the local community, but there is much more interaction with the state legislature and the federal government. One negative result of funding K–12 through property taxes is obvious: rich counties can afford to pay higher salaries and build better facilities. Thus, the poor stay poor and the rich stay rich. Bright students in poor counties are often desperate to flee to other places where the salaries are higher, with the concomitant result of the poor counties being unable to attract new businesses, thereby remaining poor.

So far, I have been attempting to explore the complex interaction the legal and educational systems have with the communities they serve. The point is how they contribute to creating the technological infrastructure of society. One final example should seal the point that our social institutions do not stand alone.

The food industry is composed of many parts. There are the producers, the consumers, the transportation systems from trucks to trains to airplanes. There is also the distribution system from companies who buy from the farmers to the large food chains to the corner grocery store. Farmers use a lot of heavy equipment such as tractors, pickup trucks, balers, reapers, and sorters, depending on the type of farming. Dairy farms need milking machines and lots of pasture. Each time a machine is involved, there is a company manufacturing that machine. The manufacture of these machines requires raw materials such as steel, which has to be extracted from the ground as iron and processed and then transported to the manufacturer who has a research and development department and a design department. The members of those departments require, often, special training in engineering and design. Large chain grocery stores are heavily computerized, making keeping track of inventory more efficient. This connects them to the computer industry. Then there is marketing and sales. Food is also important in K–12 schools, where subsidized food programs are often the only source of food for children from disadvantaged backgrounds. Associated with the consumers is the production of kitchen wares, cooking materials, cookbooks, and TV cooking shows. Cooking requires a stove and some source of fuel, electricity, or gas. The production of the fuel is an industry in itself. Further is the production and sales of materials like kitchen tables and chairs. The production of food in the form of grains is heavily dependent on the weather, and so keeping track of weather forecasts takes us to radio and television and what they require to provide those forecasts. In short, it would probably not be an overstatement to claim that the food industry permeates every facet of our society.

The production and distribution of food intersect with education and legal institutions as well since food quality is guaranteed by laws, rules, and regulations at all levels of government. To get a sense of the interrelatedness of all of this, let us take a brief look at what happened when in 2018 the U.S. federal government was shut down for political reasons. Over 800,000 people were out of work. The impact on the economy was more than significant. No salaries meant no food, which impacted the food chain. It also meant no mortgage payments, which affected the banking system. It also meant no fuel for the family car, so mom couldn't take her daughter to school or get to work herself, which impacted whatever industry she was involved with. In Washington, D.C., the impact on the restaurant business was enormous as government workers could not go out to lunch, and waiters, cooks, and clerks weren't needed. Tax returns were not being processed.

As we can see, there is considerable interaction among the various segments of our social world. The government shutdown also affected the worlds of science and technology. Among other things, the processing of grant proposals at federal granting agencies was curtailed. Research schedules were thrown into chaos as NASA was shut down and long-planned projects were put on hold. But that is not the end of the impact. Getting back up and running also take time. It is not as if everyone laid off could walk back into the office and pick up where they had stopped. There is a backlog of communications to be processed and put in the queue. Machines that were shut down have to be inspected and carefully restarted. This shutdown is an excellent example of the interrelatedness of things. It also shows, in the modern world, how complicated it all is and how tied up with our technologies we are. And it is getting more complicated as our technologies become more complicated and develop capacities to develop and run themselves, for example, self-driving cars, now trucks. Surely, buses, planes, and ships are next. And yet, with all the new technologies, it seems that we have less free time than ever. Going out for a cup of coffee with a friend during the week no longer seems doable.

Whether or not we realize it, our technologies are continuous with our lives. It is very difficult, if not impossible, to say where one feature of our lives starts and another stops. Consider the electric grid. You wake up in the morning and turn the lights, plug in the coffee pot, take a shower (lights on in the bathroom), get dressed, and go to work. When you walk into the lab, you turn the lights on and then the computer. The reach of the electric grid is very long. The role of electricity in our culture is ubiquitous and, until something goes wrong and the power goes out, transparent. There are other such technologies that make our lives better, yet we fail to look at the extent to which we rely on them. The automotive system is one. Even as it evolves from gasoline-powered vehicles to electric, from person steered to autonomous, and we debate the new changes, we still don't fully comprehend the extent to which we need this system. And it is a system. It is not just the cars. It is also the roads, the gasoline stations, and increasingly the charging stations. It includes the production of the fuels and the machinery we need to do that. The transportation of the fuels. The manufacture of the automobiles and their transportation to wherever they will be sold. The automobile dealers and the repair shops and the salesmen and the mechanics and the machinery in the repair shops, and the list goes on.

There are those who delight in "getting away from it all" leaving the complex of technological infrastructures behind them. They head for the national park for a week of camping. The hiking trails are carefully maintained. The camper has a backpack, some utensils, cooking stuff, maybe some prepared foods, $200 hiking boots, special hat and clothes, and so on. This is not getting away from it all, it is taking it with you!

It is hard to imagine one thing we do that does not involve a technological infrastructure of vast complexity. In an important sense, we are our technologies. We may not think about what is behind our being able to hit a switch and turn on the lights in our house. But that switch is connected to electric wire—which needs to be built, marketed, transported—that is connected to a fuse box—which needs to be built, marketed, transported—that is connected by miles of wire either buried or hung on poles, which is connected to a power-generating station powered by possibly coal, which has to be dug out of the ground, processed, and transported to the plant, or by a solar array or by oil or gas, which have to be extracted and processed and transported. And if the power goes out, a city dweller is helpless. No power, no lights, no elevators, no street lights, so no buses running. No sense going to work because there will be no lights, no power for the computer, you can't make coffee.

Now if we had to characterize the technological infrastructure of our society, on the surface it would appear to be a monumental task. But if we break up the societal infrastructure into smaller infrastructures, and see how they are interconnected, the job becomes manageable. We saw how the legal system, in all its complexity, interacts with the educational system and the tax system and the food system. We can now add the power system to that and the transportation system. There are several other systems of equal complexity that we don't think about that are woven into this complex. Take, for example, the clothing industry. You search into your closet for a shirt and a pair of pants. Where did they come from? Well, I bought them in a store. OK, but the store needed to be built—construction industry connection, which involves a whole other complex system involving contractors, materials, building codes, architects, and others. Returning to the store, the clothes in the store need to be designed and then made, usually in a large factory—which needed to be built—staffed by people running sewing machines (how did they get built?)—then the designs need to be displayed—orders placed, materials transported, displays in the store need to be designed and built. There need to be salespeople and cash registers and wrapping materials and hangers and display cases. It is a complex world out there, and it is of our making. But seeing it as a complex of complexes makes it manageable—we just have to ferret out the connections.

Chapter 11

A Heraclitian Philosophy of Enquiry

What do we want out of our technologies and our sciences? The answer is: control. We continue to develop new gadgets and develop our knowledge of the way the universe works so that we may control events that get in our way. And we do so despite the fact that we do not know where we are going. Thus, we ask ourselves what we need to control the situation we are investigating right now, rarely considering the law of unintended consequences. And the answer turns out to be a technological infrastructure, even if we don't know to call it that. We don't just grab things off the shelf willy-nilly. If we are trying to solve a particular scientific problem, we will probably work backward, much like you would when solving a logic poof. What do I need to get this result? How do I get what I need to get that result? And the interesting part of this process is that you are asking yourself what types of technologies you need. And yes, logic is a technology—it is a means for testing the validity of our thinking.

A Heraclitian philosophy of science and technology emphasizes the importance of the technological infrastructures required for scientific inquiry. Take the example of KATRIN—the experiment to find the weight of neutrinos at the Karlsruhe Institute of Technology. Notice first that KATRIN is taking place at an institute of technology. In an article in *Science* (Cho, (2017), pp. 1322–1327), the description of the experiment is fascinating, as much for what it fails to say as for what it does say. Neutrinos are everywhere, but

> because the particles interact with matter so feebly . . . spotting just a few of them requires a detector weighing many tons. There's no simple way to weigh a neutrino.
>
> Instead, for the past 70 years, physicists have tried to infer the neutrino's mass by studying a particular nuclear decay from which the particle emerges—the

> beta decay of tritium. Time and again, these experiments have set only upper limits on the neutrino's mass. KATRIN may be physicists last, best hope to measure it—at least without a revolutionary new technology. (p. 1322)

The fascinating point here is the acknowledgment of the reliance on technology. Should the experiment fail, the need for a "revolutionary new technology" is also acknowledged. But the point of the experiment is to solve a scientific problem, determine the mass of a neutrino. Here we have a case where not only is the science embedded within a technological infrastructure, but that if the experiment fails, a totally new technological infrastructure may be needed because a new scientific account will be required and they don't even know what it will involve. The blasé acceptance of the role of the technologies coupled with the determination to make the scientific point the important issue shows how the ideology of science corrupts our understanding of what is actually going on.

My insistence on not just acknowledging the technological infrastructure so crucial to modern science, but making it the central factor in the successes of that science, is the result of trying to find out how this process called "science" actually works. We can't continue talking about the evidence we have for a particular outcome unless we know how that evidence was produced. The process by which evidence is produced is as much a part of the evidence as the data produced at the end. In short, the evidence would not be the evidence it is without that particular process that produced it. So, the evidence is only as strong as the reliability of the process that produced it. The technological infrastructure is an essential part of that process.

The normative imperative of a Heraclitian philosophy of science and technology requires that we discover the technological infrastructure in which a science is practiced. Thus, it must be a philosophy of science *and* technology. There can be no adequate philosophy of science where the science is an abstract concept not bound to any material, social, and historical constraints, especially where those constraints are essential to the idealized practices.

Unlike a Heraclitian philosophy of science and technology, a Heraclitian philosophy of *technology* is not tied to science specifically. It acknowledges that a technological infrastructure *of some kind* is crucial to virtually any human activity not controlled by strictly biological factors. Even just sitting under a tree thinking about something requires a technological infrastructure, namely language. You can't think without language. If I ask you what you are thinking about and you can't tell me, you were not thinking. And as we noted in chapter 10, if you are going for a hike, you need hiking boots, probably a backpack to carry food, water, a first aid kit. A hat is a good idea as well and, of course, some kind of clothing and a walking stick. You can and probably should leave your phone at home unless you anticipate some really dangerous adventure, anticipating the need to call for help.

The anticipated negative reaction to the observation that we need structured technologies to do what we do will hover somewhere around the free will problem and human autonomy. But those are red herrings. We make things to assist us as we make our way around in the world. We still have free will. We can decide to leave our phone in the car when we go out for a romantic dinner. You don't *have* to use a particular technology or situate yourself in a specific technological infrastructure. But the odds are that you need to wrap yourself in some technological infrastructure or other if you are going to get anything done. You just haven't thought about it that way before. As for autonomy—no man is an island.

This brings us back to a reflection on my old definition, "Technology is humanity at work." It obviously needs to be rethought. In the previous pages, I have been laboring to explain how a technological infrastructure is essential to do science, whatever that may turn out to be. However, there is a more general point to be made here. As just noted, you need to engage some technological infrastructure to do almost anything. Perhaps you don't need one if you are a nudist and you are going for a walk. But for just about anything else, you need one. And if we consider the history of our species, it can be characterized in terms of the construction of ever more complicated technological infrastructures.

Perhaps the single most important technological invention was hay. By storing feed for livestock for the winter, we were able to abandon a nomadic form of life and settle down in one place, building first villages, then forts, then cities, then city-states, then states, nations, and so on. Along with that came beer and wine and the rest of benefits of civilization.[1] Not being on the move all the time made it possible for the various aspects of our cultures to develop: government, churches, markets, money, trade, philosophy, and others. Those aspects of our culture were themselves technological infrastructures. Why "technological" and not just "infrastructures"? Because they enable human activity, from my definition of a technological infrastructure. Yes, but philosophy? Really, how can you conceive of philosophy as a technological infrastructure? It is one because it provides the techniques and criteria for critical inquiry regarding our place in the world. It *enables* us to find our way.

To restate a trivial but profound truth: today we live in a complicated, interconnected world. We have the internet, which makes worldwide communication, among other things, possible. Through the internet, we can form a research group consisting of members on six different continents. We have a highly entangled worldwide economy. We share music, film, literature, and other forms of entertainment, either online or by traveling to various locals or by having the entertainers travel to us. And this is not even to begin to address our moves into space: satellites, trips to the moon, soon Mars, probes to the outer limits of the solar system, and now beyond. All of this because of the invention of hay.

In short, we are what we are because of what we have been able to build to enable us to do more. There is no need to wrap it all in the language of progress and rationality. The apparent order we see may be an illusion. It may be a function of our biological need to pull order out of chaos. It may also be a function of extreme hubris. We really do think we are special, and we tend to look around at what we have built and done and say "wow! no other species has done that!" But the key here is our ability to see a need for something to allow us to be something else and then get it done. So, the rethink of the definition of "technology" is "Technological infrastructures make humanity at work possible." The Heraclitian enquirer accepts the chaos—maybe even revels in it—and seeks to find some order, if only temporarily. But isn't that what we all do all the time?

I do believe we have become inured to the effects of rapid technological innovation on the way we live. It used to be the case that walking across campus was a fun experience, saying "hi" to friends, listening to students having joyous conversation, and laughing. Today it is quite different. Everyone has their heads in a cell phone—no one raises it up to see if someone they know is walking by. It has become strangely *quiet*. Likewise, two friends sitting next to each other at a bar texting *each other!* We now take for granted the role of the cell phone in our lives. And yet we fail to reflect on how it has changed our lives. And the same is true for search engines like Google. Don't bother reaching for a book to look something up—google it.

I once asked my students in a philosophy of technology class what technology they would be willing to give up. The smart-ass philosophy major in the front row answered "my toaster." The thoughtful engineering student in the back replied "my cell phone," eliciting a series of gasps from the other students. When I asked him why, he told me "I have no privacy. My phone is constantly ringing—I am always being texted and when I don't answer immediately my friends think I am mad at them and pester me all the more." That young man had thought about the effects of a technology on his life.

The bottom line is that we are what we are because of the technological systems we have created to help us do what we want/need to do. And the interesting part is that not all of those systems were rationally constructed. The internet was not created out of whole cloth—it grew slowly over a 30-year period to where we are today. And with the advent of one-answer-replies from the likes of Alexa and Echo a new massive change is underway yet again. Once we saw the possibilities and the benefits of being connected we started adding onto what was possible simply because what we added sounded at the time like a good idea. What we now call social media started as a simple idea on how to connect a bunch of college kids. Today, we find it was used to attempt to undermine a national election in the United States—who saw that coming?

I can't stress enough the need to take the law of unintended consequences seriously. But it is one thing to say it and another to figure out how to do it. In short, how can we predict the future when it will be a product of increasingly interconnected complex technological infrastructures that interact with each other in ways we can't yet imagine? We have seen this happen in the short period the internet has taken to develop into what it is today. Earlier I suggested that one way to anticipate unintended consequences is to model it using the increasingly powerful computers we are building. It should be possible to build an AI that can identify possible outcomes given current processes and their trajectories, compare them, and calculate their probabilities. We could also use computer modeling to lay out the desirability of the various outcomes and rank-order them. Oh, comes the objection, but that would mean turning our future over to machines. In a certain sense, we already have. That is the sense in which we are embedded in our various technological infrastructures.

The more I think about it, the more I am forced to conclude that this way of thinking is essentially endorsing a weak form of technological determinism. We can't stop using the technologies we have. But we can, using our technologies, plot the future paths of development and steer it, correcting for anticipated undesirable developments. When we invented hay, we did indeed set ourselves on a certain course. But there are crucial junctures along that path where decisions were made to change its direction. The American Revolution was one of those.

The American Revolution forced a break in a certain kind of development—the path of colonizing and extending the power of established governments. It didn't eliminate those activities, but it reshaped them. The United States was a colonizer, and the government of the United States has grown in power and in its reach. In a sense, it helped foment a different kind of revolution, the French Revolution, where we found we could topple one form of government and replace it with another, eventually. We rejected one form of technological infrastructure, the colonial system, and the French rejected monarchy as a style of government. We can change direction, it has been done.

But it might be objected, aren't things getting more interdependent, thereby restricting our ability to make these kinds of major changes? *It depends on how important it is to change*. We have in our hands the ability to make a major shift in direction today. As a result of our growing dependence on fossil fuels, we have created a situation where we see the planet seemingly inevitably set on the path to human extinction—I speak of climate change. According to the Intergovernmental Panel on Climate Change 2018 report, we have until 2030 to significantly reduce the production of carbon dioxide or face the fact that the planet will be uninhabitable for humans by the year 2100.

The carbon dioxide production technological infrastructure is enormous and incredibly complicated. It involves first locating the relevant fuels—coal, oil, and gas. Doing this requires a set of highly developed technologies. Then there is the extraction process, again involving highly complicated technologies and trained engineers and technicians. Then there is the transporting of the fuel to processing plants using complicated and intertwined transportation processes that include pipelines, trucks, trains, and boats, which in turn require their own supporting technological infrastructures: compressor stations, roads, railway stations, and harbors. Then there are the processing plants that turn the raw fuels into usable materials. Then there are the various items that use the fuel, from coal burning- or gas burning-electricity-producing plants to automobiles. But most important, what we haven't yet talked about, is the financial systems needed to make all this happen, such as investors, from individuals to banks and governments, or the supporting legal and governmental system of laws and regulations. How do we transform that complicated system in 12 short years to something that won't kill us?

We know what to do. We know how to create alternative sources of energy to power our needs. What is missing is the will to change. There are bright spots such as the Paris Accord, the decision by France to phase out its nuclear plants, and Sweden's decision to go solar. But there are also dark spots like a U.S. president who is a climate change denier, pulling the United States out of the Paris Accord. Time is short, and the task is mighty and yet we see more drilling for oil and worries about the impact on certain local economies than concern about climate change. Montana produces 70% of the coal mined in the United States—stopping coal mining will be catastrophic for its economy. The solution here is to find something else to employ those people, not so difficult is it?

Is this situation one in which the technological infrastructure is so entrenched that we really can't do anything about it? Well, we can, but we seem to lack the will. But it is more than that. It all boils down to money or, in the immortal words of the film *The Graduate*, "Follow the money." There is a lot of money tied up in the fossil fuel business and money buys power and the lust for that money is fueled by greed. So, Horatio, the flaw is not in a lack of ability, but in our short-sighted greed. Since the Paris climate agreement was signed in late 2015, global banks have invested $1.9 trillion in fossil fuels.[2] About half of that is invested in companies building a massive expansion of fossil fuel projects at exactly the moment climate scientists tell us we have little more than a decade to reduce emissions radically.

Scientific reports have documented climate change. The time line is accelerating for us to stop it. But how do you convince people willing to invest $1.9 trillion in developing fossil fuels to invest in something else? This situation is analogous to building a technological infrastructure for what turns out to be a failed science experiment and being unwilling to admit the theory framing the experiment is flawed. That large an investment is hard to direct to other ventures simply because it is not clear the new ventures will be as profitable. But what if you can't spend those profits because we are all dead?

We know what to do: dismantle the fossil fuel industry, construct an alternate clean energy system, develop techniques for cleaning up the current mess. We probably can do it and we should or die trying.

I know this all sounds very dark and fatalistic. But the situation is dire. It also highlights the clash of values between making money and saving the planet. No one is opposed to either making money or saving the planet. It is a matter of which is more important. If we think long term, the answer is obvious. You can't spend any of that money if you are dead. So, first save the planet by developing new nontoxic sources of energy and develop the means to clean up the mess we have created. Now there is a way to make money.

What the last several chapters have been trying to articulate is the extent to which we are captives of our technological infrastructures. They not only provide the means to do what we do, but they also constrain our thinking. Texting has become so ubiquitous that the younger generation has almost forgotten how to carry on a conversation. I have even had students tell me in class, "I can't explain it, let me text you," and pull out their phone. The language of texting is changing how we communicate, which means it is changing the way we think. The natural impulse of the young is to text rather than call, using all the shortcuts and emojis you can. If you think of the meaning of terms and concepts as the sum of the inferences you can make having employed a concept, then if the way we use language is changing by virtue of the available technologies, then we are effectively changing the way we think.

So, what is so frightening about changing the way we think? It is probably only something to worry about if it makes thinking itself impossible. I have in mind here George Orwell's doublethink in his *1984* (1949). Where do you go from "Love is hate" or "War is peace"? We are seeing something like this way of "thinking" when we hear phrases like "fake news." There was a time, not too long ago, when, if you heard the expression "news" you would infer that it is the product of serious investigating and reporting the truth. "Fake news" makes that inference impossible. I don't know what my students infer from their text messages—maybe they can't tell me!

Change in the way we think is not a bad thing—it is often the means to new discoveries. When we change scientific theories, for example, we change the way we think about the world. What I am worried about is the way our technologies are bringing about changes in our lifestyles that we are not consciously aware of. Or, if we are aware of them, like climate change, why we are failing to do anything about it?

From my perspective, climate change is an existential threat. If we fail to stop the heating of the planet, we are effectively killing our grandchildren. And, no, that is not an exaggeration. Tell that to a climate denier, and the response will be somewhere in the neighborhood of "you are hysterical." Maybe so, because I am scared to death. There is that word again.

If my account of the technological infrastructure of the fossil fuel industry is correct, we need to figure out how to unravel that infrastructure without, at the same time, bringing about more harm. I think the place to start is with my earlier proposal: find something else, preferably non-lethal, for the investors in the fossil fuel industry to invest in that will give them a better return on their investments. Redirect the money and the drilling will stop, and then the rest of the infrastructure built around the availability of fossil fuels will crumble, slowly perhaps, but inevitably. If the drilling stops, other forms of energy will be needed to replace it. That is where to invest that money.

I believe that the concept of a technological infrastructure gives us a powerful tool to think about how we live and to change the ways we live, when needed. I started by focusing on what we need in order to do science, and using the notion of a technological infrastructure, began to examine the social world we live in and how indebted we are to the technological infrastructures that support our way of life. I end with a warning that some technological infrastructures may bring about the end of life itself.

NOTES

1. You need to stay in one place for a time in order to brew beer and plant vines to make wine.

2. *Banking on Climate Change: Fossil Fuel Finance Report Card,* 2019 by Rainforest Action Network, BankTrack, Indigenous Environmental Network, Sierra Club, and Oil Change International.

Bibliography

Alberts, B., Johnson, A., Lewis, J., et al. (2002), *Molecular Biology of the Cell*. 4th edition. New York: Garland Science.

Bachelard, Gaston (1934), *Le Nouvel Esprit scientifique*. Paris: Alcan.

Berg, Jeremy (2018), "Revolutionary Technologies," *Science* Vol. 361, No. 6405, p. 872.

Bohhanon, J. (2017), "The Cyberscientist," *Science* Vol. 357, No. 6346, pp. 18–21.

Bucciarelli, Louis (1994), *Designing Engineers*. Cambridge, MA: MIT Press.

Buchler, J. (2011), *Philosophical Writings of C.S. Peirce*. New York: Dover.

Bud, R., and Warner D. (eds) (1998), *Instruments of Science*. New York: Garland Publishing Company.

Butterfield, H. (1931), *The Whig Interpretation of History*. London: G. Bell.

Chaisson, Eric J. (1994), *The Hubble Wars*. New York: HarperCollins.

Cho, A. (2017), "The Unbearable Lightness of Neutrinos," *Science* Vol. 356, No. 6345, p. 20.

Cohen, I. Bernard (1987), *Revolution in Science*. Cambridge: Belknap Press.

Cooke, Elizabeth F. (2016), *Peirce's Pragmatic Theory of Inquiry*. New York: Continuum International Publishing Group.

Daube, Joseph L. (2015), "Galileo and Renaissance Art," *Smarthistory.org*, August 9, 2015. https://smarthistory.org.

Dick, Stephen J. (1973), *Discovery and Classification in Astronomy*. Cambridge: Cambridge University Press.

Dobell, C. (1932), *Antony van Leeuwenhoek and His "Little Animals."* London: John Bale, Sons & Danielsson.

Drake, Stillman (1978), *Galileo at Work; His Scientific Biography*. Chicago: University of Chicago Press.

Egerton, F. (1976), *The Renaissance Rediscovery of Linear Perspective*. New York: Harper and Row.

Feyerabend, Paul (1975), *Against Method*. New York: New Left Books.

Galilei, Galileo (1610, 1989), *Siderius Nuncius*. Translated with Introduction, Conclusion, and Notes by Albert Van Heiden. Chicago: University of Chicago Press.

——— (1615), "Letter to the Grand Duchess Christina of Tuscany." https://web.stanford.edu/~jsabol/.../Galileo-

——— (1632, 1958), *Dialogue Concerning the Two Cief World Systems.* Berkely: University of Caifornia Press.

——— (1638, 1954), *Dialogues Concerning Two New Sciences*. New York: Dover.

Galison, Peter (1997), *Image and Logic*. Chicago: University of Chicago Press.

Goodman, Nelson (1955), *Fact, Fiction and Forecast*. Cambridge, MA: Harvard University Press.

Hanson, Norwood Russell (1968), *Patterns of Discovery*. New York: Cambridge University Press.

Hawking, Stephen (1988), *A Brief History of Time*. New York: Bantam Books.

Heidegger, Martin (1954), "The Question Concerning Technology." Translated by William Lovett. In *Basic Writings,* revised and expanded edition, edited by David Krell. New York: HarperCollins.

Heinrichs, Arianne (2009), "Stains and Fluorescent Dyes," *Nature Reviews*. http://www.nature.com/milestones/milelight/full/mo;elight02.html.

Hooke, Robert (1665), *Micrographia*. London: Royal Society.

Ivins, W. M., Jr. (1946), *Art and Geometry: A Study in Space Intuitions*. Cambridge, MA: Harvard University Press.

Joyce, James (1922), *Ulysses*. Paris: Slyvia Beach.

King, Henry C. (1955), *The History of the Telescope*. Mineola, NY: Dover.

Kuhn, Thomas (1962), *The Structure of Scientific Revolutions*. Chicago: The University of Chicago Press.

Lakatos, Imre and Musgrave, Alan (1970), *Criticism and the Growth of Knowledge*. Cambridge: Cambridge University Press.

Laudan, Larry (1977), *Progress and Its Problems*. Los Angeles: University of California Press.

Merkel, Howaard (2013), *PBS News Hour on line*, September 27.

Nanotechnology: The State of Nano-Science and Its Prospects for the Next Decade. Washington, DC: U.S. Government Printing Office, 1999. United States Congress House of Representatives.

Orwell, George (1949) *1984, Signet Classics.* New York: The Penquin Group

Pedoe, Dan (1976), *Geometry and Visual Arts*. Harmondsworth: Penguin Books, Ltd.

Pitt, Joseph C. (1987), "The Autonomy of Technology," in *Technology and Responsibility*, edited by Paul Durbin. *Philosophy and Technology*, Vol. 3. New York: Columbia University Press, pp. 99–114.

——— (1992), "Transforming knowledge: Galileo, Technology and Space Science," in *The Interaction Between Science and Technology*, edited by Bart Gremmen. *Studies in Technology and Science*, Vol. 5. Waganngen: Wall Press, pp. 99–113.

——— (1992a), "Problematics in the History of Philosophy," *Synthese*, Vol. 92, No.1, pp. 117–34.

——— (1992b), *Galileo and the Book of Nature: Method Replaces Metaphysics*, Western Ontario Series in the Philosophy of Science, Vol. 50, Kluwer: Dordrecht.

——— (1995), "Discovery, Telescopes and Progress," in *New Directions in the Philosophy of Technology*, edited by Joseph C. Pitt. *Philosophy and Technology*, Vol. 11. Dordrecht: Kluwer.

——— (1999), *Thinking About Technology*. New York: Seven Bridges Press.
——— (2001), "The Dilemma of Case Studies," *Perspectives on Science; Historical, Philosophical, Social,* Vol. 9, No. 4, pp. 373–82.
——— (2007), "Voir la Nature," in *Conceptions de la Science: Hier, Aujourd'Hui et Demain*, edited by R.M. Burian and J. Gayon. Brussells: Ousia, pp. 272–289.
——— (2016), *Doing Philosophy of Technology*. Dordrecht: Springer.
——— (2016b), "The Role of Technologies in Undermining the Perennial Philosophy," in *Philosophy and Engineering; Exploring Boundaries, Expanding Connections*, edited by Diane Michelfelder, Byron Newberry and Qin Zhu, Dordrecht. Springer, pp. 73–84.
Popper, Karl (1959), *The Logic of Scientific Discovery*. London: Hutchinson.
Price, Derek J. Desola (1963), *Little Science, Big Science*. New York: Columbia University Press.
Quine, Willard van Orman (1951), "Two Dogmas of Empiricism," *Philosophical Review* Vol. 60, pp. 20–43.
Rescher, Nicholas (1978), *Scientific Progress*. Oxford: Basil Blackwell, 178 pages.
——— (2016), *Concept Auditing*. New York: Lexington Books.
Rudner, Richard (1953), "The Scientist *qua* Scientist Makes Value Judgements," *Philosophy of Science* Vol. 20, No. 1, pp. 1–6.
Schubert, Glendon (1965), *Judicial Policy-Making*. Glenview: Scott Foreman and Company.
sciencehowthingswork.com. http//:sciencehowthingswork.com
Sellars, Wilfrid (1963), *Science, Perception and Reality*. Atascadero, CA: Ridgeview Publishing Company.
———. (1963), *Science, Perception and Reality*. New York: Routledge and Kegan Paul.
Smith, Robert (1989), *The Space Telescope*. New York: Cambridge University Press.
Sokol, Joshua (2017), "Machines That Make Sense of the Sky," *Science* 7 July Vol. 357, No. 6346.
Turner, Gerard (1980), "Microscopical Communication," in *Essays on the History of the Microscope*. Oxford: Senecio Publishing Company, pp. 215–32.
Van Helden, Albert (1989), "The Telescope and Cosmic Dimensions," in *The General History of Astronomy*, Vol. 2A, edited by Rene Taton and Curtis Wilson. Cambridge: Cambridge University Press, pp. 106–18.
Vincenti, Walter G. (1993), *What Engineers Know and How They Know It*. Baltimore: Johns Hopkins Press.
Voosen, Paul (2017), "The AI Decectives," *Science* Vol. 357, No. 6346, pp. 22–27.
Whitehead, Alfred North (1929, 1979), *Process and Reality: An Essay in Cosmology*. New York: McMillan.
Winner, Langdon (1986), *The Whale and the Reactor*. Chicago: University of Chicago Press.

Index

About the Author

Joseph C. Pitt, aka Joe, received his AB from the College of William and Mary in 1966. He completed his MA in 1969 and his Ph.D. at the University of Western Ontario in 1972. He came to Virginia Tech in 1971. He is Professor of Philosophy and of Science and Technology Studies. During his tenure at Virginia Tech he helped create an undergraduate interdisciplinary program, Humanities Science, and Technology in 1979. In 1980 he was appointed Director of the newly created Center for the Study of Science in Society in which he helped develop a PhD program in Science and Technology Studies. In the Philosophy Department he has served as Department Head, Director of Undergraduate Studies and Director of Graduate Studies. He loves to teach and has received numerous university teaching awards and served as Chair of the university's Academy of Teaching Excellence. Joe has also published 4 books and edited/co-edited twelve others and has been the recipient of several grants. He is the founding editor of the journal, *Perspectives on Science; Historical, Philosophical and Social*. Joe also served as editor of *Techné; Research in Philosophy and Technology,* the journal of the Society for Philosophy and Technology, of which he also served a president. He is a member of Sigma Xi and a Fellow of the American Association for the Advancement of Science. He and his wife, Donna live on a lovely farm in Newport, Virginia where they tend to horse and breed and raise Irish Wolfhounds.

www.ingramcontent.com/pod-product-compliance
Ingram Content Group UK Ltd.
Pitfield, Milton Keynes, MK11 3LW, UK
UKHW012042150926
475560UK00001B/151

9 781786 612359